Roger Sproston with Paul and Jayne Marshall

Straightforward Guides

ISBN

978-1-80236-007-3

Printed by 4edge www.4edge.co.uk
Cover design by Bookworks Derby

Architects drawings and designs, including cover drawing created by Paul Gaughan Building consultants www.paul-gaughan.co.uk

Acknowledgements

The authors would like to thank the following :

Paul Gaughan and David Brotherhood who were the Building Consultants who supplied drawings and design for the self-build in this book. John O'Neil for the 3D designs.

Also, Dan Duckmanton –Electricial, Steve Scott – Joinery, Heanor Bathtime - Supply and installation. And everyone else who advised and contributed!

Paul and Jayne would like to especially thank Nicholas Birkin and his team for building our home, and for his advice and friendship over the years.

Contents

Pp

Introduction   11

Why self-build?   11

Saving money   12

Energy saving   13

Who builds their own home?   13

The role of government and the construction   14
industry in encouraging self-build.

The Right to Build   14

The Help-to-Build Scheme   16

PART 1-THE INITIAL STAGES

Ch. 1-Finding a Plot of Land For Your Self-Build   23

General   23

Brownfield sites   24

Greenfield sites   25

Using garden Space (Garden grabbing)   25

Case study-finding a plot generally   26

Buy to demolish   28

Designated areas   28

Buying land at auction   34

Different types of property auction houses   35

Those who attend auctions   35

What happens on the day?   37

What to look for when plot hunting   38

The Right to Build   39

What is a serviced plot?   40
Maintaining the Right to Build registers   40
Eligibility for the Right to Build   41
Local connection   41
Accessing the register   42
    Can councils opt out of the Right to Build scheme?   42
What information do I need to supply?   42
The legal side of buying land   44
Steps in the conveyancing process   46

Ch. 2-Costs of Self-Build   **51**
Working out how much you are going to need   51
Working out the overall costs of a project   51
Purchase of a plot   52
Stamp Duty Land Tax   53
Costs of clearing a site and demolition   54
Financing costs   55
Payment of professional fees   56
Fees for planning approval and building regulation   56
Building materials and construction (labour) costs   56
Example build costs using Build Costs Calculator   58
Value added tax (VAT)   59
Reclaiming VAT   60
Eligibility   60
New homes   60
Conversions   61
Building materials   61
What doesn't qualify   61

How to claim   62

How long it takes   63

Ch. 3-Obtaining a Self-Build Mortgage   **65**

Make a detailed plan   66

Difference between an ordinary mortgage and a   67
self-build mortgage?

Payment of the stage payments   67

The arrears-type self-build mortgage   68

The advance stage payment mortgage   68

When are the fees due?   68

Example mortgage costs   69

PART 2-GETTING A PROJECT OFF THE GROUND

Ch. 4-The Design process   **75**

Designing Your Home   75

Different routes to house design-how to go about it   76

Using professionals   76

Do-It-Yourself design   77

Bespoke designs   77

Golden rules when designing a home   78

Ch. 5-Obtaining Planning Permission   **93**

Planning permission generally   93

Permitted Development Rights   95

Outline and Full planning permission   96

Costs of applying for planning permission   96

Planning application requirements 97

Design and Access Statements 97

Planning Conditions 97

Planning decisions 91

Timeframe for planning approval 99

Planning application refused 100

Planning application approved and 100
commencement of work

Carrying out works without planning permission 101

Extending planning permission 101

Example drawings for submission 103

Summary of the planning process 109

Ch. 6-Role of local Authority Building Control **113**

Building control generally 113

More about procedures 115

PART 3. UNDERSTANDING THE CONSTRUCTION
PROCESS

Ch. 7-An Overview of the Construction Process **121**

Construction management generally 121

The Contract With Your Builder 122

The Procurement Structure 123

Scope of Work or Services 123

Contract Price 124

Payment Terms 124

Work on Site 125

Changes 125

Delays & Time Extensions 125

Completion & Defects Liability 126

Liquidated Damages 127

Insurance & Liability 127

Ownership & Risk 128

Termination & Suspension 128

Longer Term Risks 129

Disputes 130

Ch. 8-Main Trades People Involved in Construction 133

Builders and subcontractors 133

Ground workers 133

Bricklayers 134

Carpenters 135

Plasterers 136

Plumbers 137

Electricians 137

Painters and decorators 138

Wall and floor tilers 138

Roofers 138

Ch. 9-Estimating Building Works 141

General 141

Case study-estimating build costs 142

Ch. 10- Stages of Construction 145

Choosing and Buying Your Site 145

Buying and Planning materials and labour 146

Overview of the construction process
with illustrations   146
Snagging   156
The Defects Period   156

PART 4. POST CONSTRUCTION AND MOVING INTO YOUR HOME

Ch. 11-Steps Following Completion Of Your Home   163
Completion of the build   163
Completion certificates   153
Release of final funds   164
Insurance   165
VAT reclaim   165
Retention money   165
Log book and instructions   165
Disputes   166
Mediation and conciliation   167
Arbitration   167
Litigation   168

Conclusion to the book   170

Useful addresses and websites
Index
Appendix 1. Planning consent 3 bed Dorma bungalow (Paul and Jayne Marshall)

Introduction

For many people, the idea of building their own home remains just that: a nice idea, but unfortunately not achievable. First off, there is the thought that self-build means getting down onto the building site and doing all the work your self. However, most often the reality is that self-build means that the self-builder works with a variety of people and businesses, such as architects and builders to construct their home.

Why Self-Build?

Paul and Jayne Marshall:

We first got involved in self-build in 1999, and got the idea initially through friends who were looking for properties. They found that the cost of a house at that time was far too expensive and decided to explore other avenues.

We were also in the same boat, and discussed the problem with a builder friend of ours, who offered to build a house for us, subject to finding a plot of land, which is always the main consideration. We found a plot in our home village, a house with an orchard, both on the market but later split into two plots, of which we bought one, which had been granted outline planning permission. The initial price was £40,000. We approached the estate agent as it had been on the market for three months and offered

£25,000 for the plot. This was rejected and Jayne contacted the owner who met us on site. She stated that she had never received the revised offer. We offered £27,000 which was accepted. This was our very first self-build and we have never looked back.!

On the face of it, it seems easier to just buy a house, without getting involved in the whole process. However, deciding to self-build a home and maintain control of the design and build means that you can end up with the home you want, designed to your own specifications and at a much lower cost than a traditionally bought home. Building your own home from scratch is the ultimate chance to create something that is completely designed around your own unique needs and your future aspirations.

Self-built homes can also combine the best features available to homeowners. For example, labour-saving features such as a central vacuum and even automated home technology can be incorporated at a fraction of the cost you might expect. These features can revolutionize your lifestyle: at the touch of a button you can control the heating, climate, lighting and entertainment functions of your entire home.

Saving money

By going through the process of creating a design and having a personal input with the house building process,

12

you can usually save yourself 30% plus on the market value. As a guide, the house that you spend £200,000 (on the full project) should quite easily be worth £300,000 plus if all goes to plan. The potential equity to be made also means that many self-builders repeat the process three or four times, ending up with a bespoke home that is mortgage free.

Energy saving

One other big advantage of self-build is that you can design your home to be green, sustainable and energy saving. A green home is designed to be highly efficient and to make use of natural energy harvested from the local environment. The majority of self-builds today incorporate some green features and as a result, their energy profile and carbon footprint can be tiny when compared to a conventional house. This means that not only is the effect you have on the environment reduced, but it is likely that the effect energy bills have on your wallet will be reduced as well.

Who builds their own home?

All types of people choose to build their own home – from young couples and singles starting out on the property ladder to families and those wanting a retirement pad for two. Age is certainly no barrier, whilst for those with accessibility needs, and their families, self-build is often the only way to achieve a home to fit their special requirements.

Finance and DIY skills (or a lack of) should not be deterrents either. Self-build can offer many people on low incomes a chance to own their own home without having to fork out local market prices — custom and community build projects are a fine example of this (see chapter 1).

So, where do you start? This book covers all the essential stages of self-build from finding land and finance, the design process, obtaining planning permission, the construction process and the final process of moving in.

As a backdrop to all this activity, it is very important to understand the moves that have been underway to promote self-build, making the whole process easier, both from national organizations such as the National Custom and Self-Build Association (NaCSBA) and also within government.

The role of government and the construction industry in encouraging self-build.

One of the biggest historic problems facing those with a passion for building their own home has been the difficulty in finding land and obtaining finance to get the project off the ground. During the last five years, pressure from organisations such as the above mentioned National Custom and Self-Build Association (NaCSBA) has been brought to bear on Government to ensure that the process of accessing land and obtaining finance to self-build a home becomes a lot clearer and a lot simpler. This has culminated in the Right to Build.

The Right to Build

This is a scheme designed to help combat the housing crisis and get more people building their own homes. In a bid to boost house building and address concerns about affordability, the Government has passed legislation to make more individual and serviced plots available for those wanting to build their own homes.

The Self-Build and Custom Housebuilding Act 2015, is an Act which places a duty on local authorities to keep a register of individuals and community groups who have expressed an interest in acquiring land to bring forward self-build and custom-build projects and to take account of and make provision for the interests of those on such registers in developing their housing initiatives and their local plans. It also allows volume house builders to include self-build and custom-build projects as contributing towards their affordable housing obligations, when in partnership for this purpose with a Registered Social Landlord.

Going further, as part of the Housing and Planning Act 2016, which came into force at the end of October 2016, local authorities are now required to help find land for those who have an interest in building their own home.

This is done through *the Right to Build register*. Anyone who is interested in building their own home is invited to register their interest on the Right to Build Portal (a site set up by the National Custom and Self Build Association (NaCSBA).

The scheme currently extends to England only, with discussions underway with the Welsh and Scottish assemblies. In Northern Ireland, land supply is less of an issue, with 15% of new homes being commissioned by their owners.

In Chapter 1, we will be discussing the Right to Build in more depth.

The Help to Build Scheme

Under the Help to Build Scheme, active from November 2021, you can spend up to £600,000 to create the exact house you want with a deposit of only 5 per cent and an interest free (for five years) government loan, under rules of the Help-to-Build scheme. Anyone in England can apply for funding for a self-build home, in which you find a plot and manage builders yourself, or a custom-build home, in which is overseen by a developer to your specification. The government will offer loans of between 5 per cent and 20 per cent (up to 40 per cent in London) of the land and build costs, capped at £600,000. To fund the rest of the cost you will need a cash deposit of at least 5 per cent and a self-build mortgage from a lender registered with Help to Build.

The £600,000 Help to Build limit is enough for the plot and construction of a four-bedroom house in a two-thirds of England, according to exclusive analysis by the property portal Zoopla for the National Custom and Self Build Association (NaCSBA). Although it won't stretch to that across most of London, its commuter belt, the Cotswolds and parts of the south coast.

So would you qualify for Help to Build, and how much could you really borrow? Speak informally to a self-build mortgage broker such as Buildstore or a provider such as Ecology Building Society to find out. Self-build mortgages typically cover as much as 85 per cent of the land and build cost, capped at up to four and a half times your (joint) gross income. To get a mortgage you must have planning permission on the site. A Help to Build loan reduces the size of the deposit you need, which would otherwise be at least 15 per cent. There is one further test that says the build must cost no more than £400,000, which is plenty to build a four-bedroom home. This is meant to stop too-large homes where the land is already owned. Help to Build is more generous than Help to Buy, the equivalent scheme on new homes bought from developers, in that existing homeowners are eligible and there are no regional price gaps. It means people with small deposits no longer have to choose a home that someone else has decided they will like.

17

How does Help to Build work?

After you are offered an equity loan, you'll have three years to build. Your self-build lender will release funds in stages as the build progresses. Once your home is complete the government will pay the Help to Build equity loan amount to the lender and your self-build mortgage will automatically switch to a repayment mortgage. The Help to Build loan "is not a discount" on the land and build cost, the scheme's prospectus warns. The government will start charging you 1.75 per cent interest on its portion from year six, rising annually. You must repay the equity loan at the end of the term (normally 25 years), when you sell the house, pay off your mortgage – or you can repay it at any time before. The repayment sum is linked to the value of your home at the time, not the amount you originally borrowed. The first £150 million Help to Build pot could fund 4,000 to 5000 homes. Applications will open in winter 2021.

In Wales the £210 million Self Build Wales scheme lists plots and offers government loans of up to 75 per cent of the land and the full build cost. Meanwhile eariier this year Scotland's self-build loan fund was extended and now has a closing date of August 31 2022. Applicants can apply for loans of up to £175,000 to help with construction fees on self-build projects (mygov.scot).

For more details of Help to Build go to:

https://www.gov.uk/government/publications/self-and-custom-build-action-plan

A word about the authors:
Roger Sproston BA MSC is a former housing manager and developer with many years of experience of working with self-build groups within the co-operative movement.

Paul and Jayne Marshall have been involved in self-build for a number of years and have completed three projects. They bring an enormous amount of practical experience to this book, which will be illustrated through case studies.

****

PART 1

THE INITIAL STAGES

- Finding a Plot of Land

- Working out the Costs

- Obtaining Finance

Chapter 1

Finding a Plot of Land for your Self-Build

Around 13,000 people successfully self-build every year, so clearly the plots are out there. As stated, it is the government's intention to dramatically increase that number. However, unlike the normal housing market, individual building plots are not so obvious to find – you have to work that much harder to secure a good one.

In some cases, it can take years to find a plot – especially if you're particular about elements such as size and the amount of work you're willing to take on. So be prepared to revise your goals if your search isn't going well. Flexibility, combined with the ability to focus your time and energy on the hunt, will give you the best chance of success.

There are many routes to finding a plot. As mentioned in the introduction, local authorities are now obliged to help those seeking land for self-build through the setting up of a Right to Build Register in their area. We will be discussing this in more depth at the end of this section.

One thing is true-when you are searching for a building plot you are up against some heavyweight competition, such as other self –builders, full-time 'land finders', such as small builders and all the vested interests of the development industry who are continually looking for viable plots. However, small self-builders now have government on their side. Before you embark on your search it is useful to understand the type of plots available and the ones that are restricted. This affects the possibility of planning permission.

Brownfield sites

The government focuses much new housing on brownfield sites (previously developed land), so local councils should look favourably on plans for these plots. Services are likely to be already in place, too. However, you'll need to apply for a change of use, and design restrictions may be imposed, such as maintaining the previous building's footprint. Brownfield sites are usually relatively affordable upfront, but you may need to factor in costs for a buy to demolish project (see below).

Greenfield sites

This term refers to land that's not been built on before – whether open countryside, gaps in rural areas, on the outskirts of villages or between existing houses. It's not impossible to gain planning permission to build on a greenfield site, but there's a distinction when it comes to fiercely-protected 'Green Belt'. Opportunities for an entirely new home in green belt are rare – you're more likely to be granted permission for an extension or buy to demolish scheme.

Using garden Space (Garden grabbing)

The definition of brownfield land has been extended to exclude domestic gardens. Nevertheless, Planning Policy Statement PPS3-Housing-still advises that 'options for accommodating new housing growth may include additional housing in established residential areas'. So infilling and small scale development on gardens is still possible, and one-off houses are likely to be preferred over compact development – so self-builders contemplating this approach can breathe a sigh of relief!

Case study-finding a plot generally

Paul and Jayne Marshall

Although there are numerous ways to source plots, for example through estate agents and also online, we found ours through word of mouth. In Derbyshire, this is perhaps easier than in the south, as there is more available land here. Our second self-build, Jubilee Court, one of ten plots, we found through a friend and we bought it with outline planning permission for £110,000. Jayne designed the house and put detailed planning and landscaped the plot, then we sold it on for £135,000 in 2012.

We found our third plot by approaching a family that we knew. She had a large plot, used as an allotment which historically had 4 cottages on it (overleaf).

Three years earlier she had been approached by a local estate agent who, with an architect put in for detailed planning permission. This was initially refused on traffic grounds and as a result of the findings of the topographical survey, the architect informed the owner that significant sums would need to be spent, and the owner got cold feet. Jayne had been following the progress and approached the family with an offer of £90,000 subject to full planning permission, which would cost £7,000 paid for by Jayne. Six months later, full planning permission was granted.

In our case, we have found the building plots through word of mouth. We also recommend looking around the area that you are interested in for potential plots. They don't have to be whole plots, they can be large gardens, side accesses or small areas of disused land. Searching online can also be fruitful. However, one major consideration is where you live, what part of the country and the overall demand for land.

If you are keen to embark on self-build as an option then in the end you will find your plot!

Buy to demolish

This is popular because it's usually cheaper than renovating an existing property in spite of the fact that demolition fees can run into many thousands of pounds. You're less likely to encounter hidden costs by knocking down and starting afresh, and VAT is reclaimable on new-builds but not refurbishments. You'll generate a lot of waste by demolishing an existing property, but you could sell-on salvageable materials such as bricks or even re-use them yourself. You may only be allowed to build to the same height and footprint as the previous building.

Designated areas

Self-building in locations with special designations – such as conservation areas – is subject to strict controls. In these cases, you're very unlikely to be granted planning permission for a new house, or even a 'demolish and rebuild'. Renovation opportunities are a better bet, but you'll find that permitted development rights are often severely restricted.

So, bearing in mind the above, here are a few tips when looking for a site.

➢ Be flexible: one reason that people fail to purchase a plot is because they will not compromise. You need to be creative in your thinking. It is sometimes better to have an architect on board when appraising a plot to help you visualize what you can build on a plot and also to advise on the possibility of obtaining planning permission.

➢ Following on from the above, it helps to understand the different types of plot available. When you are trying to find the ideal plot, it often helps to think outside the box. Often pieces of land are ripe for building a home on, but not advertised as such. It therefore helps to understand the many different types of building plot which may be available to you, if you are willing to look that little bit harder. For example, there may be a small plot sitting behind some existing house, not visible from the street. And don't overlook houses that are for sale that may not be what you are looking for, but could be demolished and replaced with a home more suitable.

➢ Use maps: Using Google Maps and even Streetview is a big help to the would-be self-builder looking for plots. You'll be able to identify gaps in the

29

streetscene, small bungalows on large bits of land, and potential backland plots, all of which are ripe for redevelopment. Following on from this, familiarize yourself with an area and gather as much information as possible. Define your area and don't pick too large an area.

➢ Use websites: there are a number of specialist 'plot finding' websites that you could investigate. Collectively, they list thousands of self-build plots in the UK and offer a powerful way of searching and contacting listed vendors.

They include:

o PlotBrowser
o Plotfinder.net
o PlotSearch
o Gumtree (Gumtree.com)

Additionally, there are other online sources that sometimes list self-build plots, including:

o Prime Location
o Rightmove,
o Zoopla
o Movehut.

Plus, there's The Land Bank Partnership; a useful site which specialises in the sale of land with a planning consent or the potential for residential development in the West/South West of England.

➢ Visit planning departments: always a good idea because if anyone wishes to get planning approval to build on a piece of land, they must first submit an application (see chapter 5), which then becomes a matter of public record. You can walk into any planning department and ask to see the Planning Register, where all the applications and decisions (where they have been reached) are recorded. Many council's publish them on their websites. What you are looking for are recent applications, preferably outline (i.e. no detailed drawings), for single houses. If an approval has not come through, so much the better. A plot will not usually be advertised for sale until the planning approval has been granted, because this enhances the value, and, if someone spots it early enough, they can make an approach before many others are even aware that it is going to be for sale.

➢ If you find such an application, note the applicant's details and approach them directly. If the application is for outline approval there is a good chance that they are planning to sell the land, because there is no point in getting a detailed set of plans drawn up which may be changed by a purchaser. But sometimes they may have obtained detailed approval, with a full design, probably because the planners have insisted on it. Whatever the situation there is no reason why you should not make a polite approach, either by letter or telephone.

➢ Estate agents: register your interest with estate agents (especially independents) in the area you're keen on. Get in touch with local surveyors and architects too, as they'll find out about new plots early. However, you should understand that not all estate agents will be open with you-their job is to make as much as possible and they may want to reserve plots for their favoured clients. The first thing to do is to register with agents in the areas that you are looking for.

➢ Look out for custom build schemes in your area: many plots are beginning to come to market through the custom build route, whereby developers and councils release land for large-scale self build. These may be a handful of serviced plots on the edge of a new development, or a new community of self built homes planned by a council. Check out plotfinder.net for details as they emerge, and also the Government's main self build information site.

➢ Make enquiries of builders: although builders are usually in competition with self-builders, there are some circumstances in which they may want to help you. Sometimes a small builder will not want the risk of developing a site, perhaps because of cash-flow problems, and may be prepared to sell you something from their 'land bank'. They sometimes however, add conditions, such as that you have to use them to build the new house already designed by them. This can be a significant drawback, because if you agree to it before you have detailed plans and specifications you may find that the construction cost is very high. Make sure that you know what you are getting into first

➢ **Beware crooks**:: there are people who run property scans and are fully prepared to exploit plot hunters and relieve them of their money. This has been well publicized. These companies offer what are apparently prime potential plots, for a bargain price. The catch is that there is no planning approval. It is suggested that, in the fullness of time, the land may eventually get planning approval, and you will then own a prime building plot. The truth is usually that it probably never will, and you have wasted your money. The main point here is to avoid these people like the plague. Never get involved-do so at your own peril! The website PropertySCAM lists many of the ongoing land scams where people are being conned. Also:

Https://www.consumerfraudreporting.org/current_top_10_scam_list.php is a very useful site.

> ➤ Using self-build companies and architects: there are a few companies, some connected to kit suppliers, who buy up larger sites, split them into individual properties, and sell them on to self-builders. You should always check whether you are tied into using a particular firm if you buy a plot. If this is the only way you can get a site in the right area, make sure that you get independent advice before entering in to an agreement.

Buying land at auction

Property auctions are another avenue for finding land for a self-build. The process is very similar to the normal method of private sale. However, for an auction sale the seller and their solicitor carry out all the necessary paperwork and legal investigations prior to the auction. Subject to the property receiving an acceptable bid, the property will be 'sold' on auction day with a legally binding exchange of contracts and a fixed completion date.

Different types of property auction houses

Auction houses vary in size and the amount of business that they conduct and the frequency with which they hold auctions. Most will sell both residential and commercial property and each will have its own style of operation, and fee structure. Large auction houses will hold auctions frequently, perhaps every two months and will have around 250 lots for sale. A lot of the auctions happen in London and the main cities or large towns.

Most of the large auction houses will deal with property put forward by large institutions, such as banks selling repossessions and also local authorities and will advertise the sales in the mainstream media and trade papers. The medium size auction houses will hold auctions as frequently as they can, in regional venues, such as racecourses and conference centres, and depending on stock, usually every two to three months, tending to advertise locally.

The small auction houses will have far fewer lots and will hold their sales in smaller local venues. They may advertise in local press but more often will trade on word of mouth.

Those who attend auctions

As you might imagine, all sorts of people attend auctions. The common denominator is that they are all interested in buying property.

Property investors are most common at auction, people who are starting out building a portfolio or those who have large portfolios that they wish to expand. They tend to fall into two groups, those who are after capital appreciation, i.e. buy at a low value and build the capital value and those who are looking for rental income. Then there are the property traders who like a quick profit from buying and 'flipping' property. These types usually have intimate knowledge of an area and are well placed to make a quick profit.

Then we have the developers and the self-builders (you) who look for small sites or, in the case of developers, larger sites where property can be built and sold on. The sites can have existing buildings on them or can be vacant lots with or without planning permission.

Although a wide cross-section of properties are to be found at auction, for the budding self-builder it will either be building land, land with existing property for improvement or demolition or development propositions such as derelict or disused farm buildings, buildings with potential for conversion or change of use.

The latest date for entering property for an auction is usually five to six weeks prior to the auction. Once the marketing agreement has been signed, the property will be placed in the catalogue and a board erected. Each seller's legal representative will be contacted to obtain a legal pack, which the seller must produce. This pack should generally include office copy entries and plans, the local search, leases (if applicable) and any other relevant documents. All properties at auction are sold under the General Conditions of Sale and, with the legal pack, also require any Special Conditions of Sale to be attached.

These are matters that are relevant solely to the lot being sold. The marketing period starts five to six weeks before a sale. The details of all the lots to be offered in the next sale, including colour photographs of each property, viewing arrangements and any other relevant information will then be published. A few days prior to the auction, the reserve price will be agreed.

What happens on the day?

The lots will be offered and the bidding taken to the highest possible level and once the gavel falls, the contracts will be exchanged. The buyer purchases the property at the price they bid - this cannot be negotiated and the stipulated terms cannot be changed. The buyer will then pay 10% of the purchase price on the day and completion occurs 28 days later. The funds are then paid to the seller less the fees of the Auctioneers and those of the seller's solicitor.

The atmosphere of an auction room can be extremely exciting and competitive and it is often the case that an interested party will bid in excess of the figure that had previously been set as their maximum. In some cases, the prices achieved at auction can be higher than those achieved by private treaty.

What to look for when plot hunting

If you are going it alone and looking for your own plot of land, then when you are out scouting an area, these are some of the clues that you should look for:

> Large gaps between and behind houses. It is usually easier to get planning approval for development in between, or next to, existing houses. If there is space beside a house, and especially if it has easy access to the road, it is a potential plot. If there is a big back garden, and access for vehicles to get to it down the side of the house, it may be possible to build at the bottom of it.

> Look for houses of a similar size and quality to the one you wish to build. The way that houses are valued means that it is less economic to develop a house that is massively disproportionate to those surrounding it. You can end up over-developing, that is spending far more money on a house than you could ever sell it for; or under-developing, that is building too small a house and failing to realize the full potential of the site.

> Vehicle access. Whatever land you find, unless it is near a city or town centre, will have to have parking space, so there must be a way of reaching it by car.

> Disused land and brownfield sites: these are very easy to miss. It takes a lot of imagination to see an

old factory or disused industrial unit, or a scrap yard as the site for a potential home, but they all could be, subject of course to the all important planning approval.

Last but not least there is the possibility of assembling your own site. For example, if you see a number of gardens that are too small for a house, but combined could be big enough, take a leaf from the professional developer's book and consider assembling your own site. It needs tact and patience, but it can be done-particularly when the homeowners realize that a small bit of their garden can earn them some money.

There is one further important way that you can potentially access a plot of land and that is through the local authority Right to Build Registers, as outlined earlier.

The Right to Build

As mentioned in the introduction to this book, on 31st October 2016, The Self-Build and Custom Housebuilding Act 2015 and the Housing and Planning Act 2016 combined to establish a right to build your own home. Essentially, this means that councils have a duty to find land and grant planning permission for sufficient *serviced plots* to meet demand, measured on so-called Right to Build registers, within a three-year period (and on an ongoing basis).

This is only for would-be self-builders and not for speculators. The measures have been introduced to encourage individuals (and small groups) to build their own homes.

What is a serviced plot?

For the purposes of the Right to Build legislation, the definition of a serviced plot is that it must have access to a public highway and connections for electricity, water and waste water (or that these can be provided in specified circumstances or within a certain timeframe). This means mains gas connections are currently excluded, presumably because there's not always a supply available in remote locations. There is provision for further services, such as broadband, to be added to the list at a later date – and of course some local authorities and landowners may choose to provide the infrastructure for these anyway.

Maintaining the Right to Build registers

Fundamentally, the bodies charged with this responsibility are the district and borough councils, unitary authorities and national park authorities in England – of which there are 336 in total. This tally excludes the 27 county councils, which don't usually deal with planning matters and are only required to keep Right to Build registers for areas that aren't covered by district bodies.

(Delegated powers mean the Self-Build and Custom Housebuilding Act doesn't automatically apply in Wales, Scotland or Northern Ireland).

The Welsh Assembly has the power to vote it in, should it decide to, but in Scotland and Northern Ireland there would have to be a separate legislative process.

Eligibility for the Right to Build

The criteria are fairly loose. As long as you're over 18 and a British citizen or national of a European Economic Area or Switzerland, you can make an application to join any Right to Build register. There's another key qualifying factor: you must be seeking to build a home in the authority's area as your sole or main residence. So this route isn't for people looking to develop a house to sell on or a buy to let.

Local connection

If an authority sets this kind of criteria, only those who meet these additional eligibility checks will be included in Part 1 of its register. Crucially, this is the list for which councils are required to grant planning permission for sufficient plots.

There's no obligation to do the same for those who qualify only for Part 2 (no local connection), although this data may be used to generate statistics about self-build and inform local policy.

Accessing the register

The vast majority of councils have chosen to run their own registers. These can generally be accessed via an online form – but in some cases you'll have to download a static form and send it in by email or post.

Can councils opt out of the Right to Build scheme?

Amendments to the Self-Build and Custom Housebuilding Act made in 2016, include a section regarding exemption from the duty to grant planning permission for enough suitable plots. This is presumably intended as a fallback for areas where there's high demand but significant challenges to land supply – such as in city centres – and the authority must apply to the Secretary of State to obtain an exemption.

What information do I need to supply?

This is fundamentally down to your local authority. Some only ask for the bare minimum the legislation requires – which means supplying your name, address and a declaration that you're eligible for inclusion on its Right to Build register (generally they'll only accept one entry per household).

The vast majority of councils will ask for some additional detail about your interest in building your own home. Common questions include:

- Whether you have a local connection to the area
- What level of involvement you want
- An idea of the size of plot you're after (in m2)
- How many bedrooms you'd like to have
- When you'd be able to start on site (typically within a year, one to two years and three years or more)
- Your preferred area
- Why you want to build your own house
- Where your finances stand (do you have savings and/or a home to sell, for instance, and will you need a self-build mortgage?)

The council should contact you within 28 days to acknowledge your application for the register and let you know whether you've been successful in getting entered onto the list for that base period. Depending on how much information was requested, you may be asked to provide further details.

You can change the details of your listing if your circumstances change – such as if you've decided to widen your search for a self-build plot to additional parts of the council's jurisdiction. However, be careful when doing this, as any changes you request that affect your eligibility (either wholesale or for the crucial Part 1 list) could see your name struck off the register.

Most local authorities don't currently charge for entry onto their registers, but the regulations that give them the power to set a fee took effect on 31st October 2017.

On the face of it, everyone on the list should be offered the opportunity to purchase a piece of land. But sites won't become instantly available – councils have up to three years to grant permission for the plots requested in each 12-month base period.

In some cases councils will only have to offer you a plot if you're eligible for Part 1 of the list. It's also worth bearing in mind there's no guarantee a given site will match your expectations (the more details you give, the better your chances).

In Chapter 5, we will be looking at the processes involved in obtaining full planning permission once you have found your desired plot. Many plots come with outline planning permission but it will be necessary to obtain full permission before you can commence your self-build.

The legal side of buying land

Buying a property, whether a plot or building, is by it's very nature, complex and you will need a solicitor to make all the enquiries to ensure that the vendor actually owns the property and is free to sell it and also to ascertain what, if any, restrictions apply to the property.

The process of buying an empty building plot is similar to buying an existing building. Much of the solicitors work will go into finding the exact conditions of (and limitations to) the use of the property, the details of the transaction itself and the type of ownership gained, whether freehold (usual) or leasehold.

Although most land is sold as freehold, some land is sold subject to a lease. During the period of the lease a ground rent has to be paid to the freeholder and there will also tend to be more restrictions placed on the use of the land. The length of the lease will also be a factor in a lender's willingness to advance money for the project.

Before you even consider buying the land, the following questions need to be asked of the seller to determine whether or not you will go ahead with the purchase:

➢ Is the property freehold or leasehold-if leasehold how long is the remaining term and how much is the ground rent?

➢ Is the access road and main drainage the responsibility of the local council, or is ownership and maintenance shared by the property owners?

➢ Does anyone have the right of access to any part of the plot or property for any reason?

> ➤ Are there any restrictions in the deeds or elsewhere to the use of the plot or property, and if so what are they?
>
> ➤ Are there any compulsory charges for maintenance or renewal on any aspect of the property, and if so what are they?
>
> ➤ Is it certain that the plot or property is to be vacant when sold?
>
> ➤ Has anyone been using the property or part of it for 12 years or more and claimed ownership of it?

Steps in the conveyancing process

If you have decided to go ahead and purchase the property then the following work will be undertaken by your solicitor to complete the deal. They are the same steps that are taken with conventional residential purchases:

> ➤ The Buyer makes an offer on the property, which is accepted by the seller.
>
> ➤ The Buyer's Conveyancer instructed on acceptance of the offer
>
> ➤ The Buyer arranges a survey on the property, and makes an application for a mortgage (if required). Note that this will have already been agreed in principle.

- The Buyer's Conveyancer confirms instructions by letter setting out the terms of business and fixed fee costs.

- The Buyer's Conveyancer contacts the seller's Conveyancer to obtain the contract pack.

- The Buyers Conveyancer checks the contract pack, raises pre-contract enquiries, carries out the necessary searches and obtains a copy of the mortgage offer.

- The Sellers's Conveyancer and seller answer pre-contract enquiries and return these to buyer's Conveyancer.

- The Buyer's Conveyancer reviews and reports to the buyer on the contents of the contract pack, pre-contract enquiries, the result of the searches and mortgage offer. The buyer then considers this report and raises questions on anything that is unclear.

- When the buyer is happy to proceed, arrangements are made for the deposit to be paid to the buyer's Conveyancer in readiness for exchange of contracts.

- The Seller and buyer agree on a completion date and contracts are formally "exchanged" - meaning both parties are legally committed to the transaction.

➢ The Buyer's Conveyancer prepares a draft transfer deed and completion information form and sends these to the seller's Conveyancer for completion.

➢ The Seller's solicitor approves the draft transfer deed and a final copy is made. This may need to be signed by the buyer before being sent to the seller's solicitor for signature by the seller in readiness for completion.

➢ The Buyer's Conveyancer prepares a completion statement, carries out pre-completion searches and applies to the buyer's mortgage lender for the mortgage loan.

➢ On completion, the buyer vacates the property by the agreed time and buyer's Conveyancer sends the proceeds of sale to the seller's Conveyancer.

➢ The Seller's Conveyancer releases the keys (or access to the plot) to the estate agent (if one was used) and sends the title deeds and transfer deed to the buyer's Conveyancer together with an undertaking to repay any existing mortgage.

➢ The Buyer's Conveyancer sends the stamp duty payable to HMRC, (if appropriate) receives the title deeds, transfer deed and proof that the seller has paid the outstanding mortgage on the property.

➢ The Buyer's Conveyancer registers the property in the name of the buyer at The Land Registry.

> ➤ The buyer receives a copy of the registered title from The Land Registry. Any documents required by the mortgage lender to be retained by them are sent on by the Buyer's solicitor.

And that's it. After finding a plot the land is yours ready to go to the next stage, which will be the design stage, getting everything in place, and obtaining full planning permission (discussed in chapter 5) before starting on site.

Before this, in the next two chapters we will discuss the costs involved in self-build and also how to obtain finance through a self-build mortgage.

Now read an overview of the main points from Chapter 1 overleaf.

****

Overview of the Main Points from Chapter 1-Finding a Plot of Land

- Around 13,000 people successfully self-build their own home every year. This is set to increase.

- Be prepared to revise your goals if your search for a plot isn't going well. Flexibility, combined with the ability to focus your time and energy on the hunt, will give you the best chance of success.

- Once you have defined the type of plot you want, then in addition to the websites mentioned in the book, there are other online sources that sometimes list self-build plots, including PrimeLocation, Rightmove, Zoopla and Movehut. Plus, there's The Land Bank Partnership. And, don't forget auctions and planning departments!

- Under the Right-to-Build legislation, councils have a duty to find land and grant planning permission for sufficient serviced plots to meet demand, measured on Right to Build registers. It is always worth investigating this route as you may qualify.

- Two more points-beware of crooks offering land. It is always a waste of time and money. Secondly, always ensure that the legal side of purchasing a plot is exhaustive and that you end up with what you want.

Chapter 2

Costs of Self-Build

Working out how much you are going to need

One of the main aims of self-building a house is to ensure that you end up with more money than you started with, in terms of the final value of the finished property. The basic principle of self-build finances is to put yourself in the shoes of a developer so that the developer's profit margin is yours.

Working out the overall costs of a project

With some projects, working out the final costs will be easier than others. You might be in the process of striking a deal with a builder whereby they will design and build the whole property. In this case, a lot of the groundwork will be done in terms of arriving at a build cost. However, in cases where you have a lot more input into the various stages, costing a project will be a little more complex.

Paul and Jayne Marshall

There are a number of ways of arriving at an overall cost for constructing a self-build property. Of course there are the costs associated with the acquisition of land, such as the cost of the land and any associated fees and then there are the all-important build costs of between £45-£65,000.

One way of estimating is to get an overall cost to design and build a property from a builder. This is perhaps the most expensive way and also takes away from the design element, particularly if you have strong ideas of your own.

We chose not to go down this route. We obtained a quote from our builder for labour only, minus electrics and plumbing. The labour only cost was £15,000 (at the time- 2009) and we agreed three stage payments with the supplier of materials (Build base) based on an estimate from the builder. By doing it this way, we had a greater choice over materials. We were able to claim back VAT on materials. One important point, our choice of materials was, in part, influenced by the requirements of building control, for example, slate for the roofing.

Listed below are some of the major items involved in self-build and also their associated costs (where known).

Purchase of a plot

The amount of money spent on a plot will depend on a number of factors. This can range from no expenditure if you are using your own back garden, for example, to hundreds of thousands of pounds, depending on what you buy and where you buy it. The cost of a building plot will obviously depend on where it is, from the cheapest areas to the most expensive (for example buying a plot in the North East of England in an urban area (£20,000)) will be a lot cheaper than buying a plot in the South East (London or Brighton for example) where you might spend up to £500,000 or more.

Stamp Duty Land Tax

A cost to watch out for is stamp duty when purchasing the land. This will depend on the cost and the threshold in force at the time.

When you buy an existing built dwelling, you're liable for stamp duty land tax (SDLT) on the entire value of the property. However, when you buy a plot of land you pay SDLT on the value of the land only and no SDLT is payable on the completed home.

In addition, vacant plots are often considered by HMRC as being liable for commercial rates of stamp duty rather than residential rates, and thus any surcharges can be avoided as this applies only to residential purchases. However, if you buy a plot that is part of a garden of an existing house, you're likely to have to pay SDLT at the residential rate.

For example, if you purchase a £400,000 property, you'll now have to pay £22,000 in stamp duty if it's a second property. But if, instead, you buy a vacant plot of land that's deemed as commercial and build a property to create something worth £400,000, you could possibly avoid stamp duty altogether. If, for example, you buy an empty plot that costs £140,000 and stamp duty is charged at commercial rates, as the purchase prices falls under the lowest threshold for non-residential properties of £150,000, you won't have to pay any stamp duty.

For a more detailed outline of stamp duty rates you should go to www.gov.uk/stamp-duty-land-tax.

Costs of clearing a site and demolition

This will be a significant cost if it is appropriate to the plot that you have purchased. If the site is a brownfield site (a site previously used for industrial or commercial purposes) or when doing a conversion or replacement dwelling you will need to consider the costs of demolition and clearance. The prices below are correct as at 2022.

The cost of demolishing a house ranges from £6,600-£8,800 for a small 80-120m2 detached property, up to £13,200-£16,500 for a more substantial property of 200-250m2. But costs will be mitigated if there are materials with salvage value. A demolition contractor (find one at demolition-nfdc.com) will list the value of any salvage items and offset this against the quote for demolition work and site clearance.

Existing service connections to electricity, mains water and sewerage, and connection to the highway can usually be reused, which can represent a significant saving compared to developing a virgin plot.

A large part of the cost of demolition is landfill and haulage, so costs can be reduced if there is scope to reuse or dispose of non-toxic waste on site — such as clean hardcore for drives, paths, terraces and soakaways.

The time taken for demolition work depends on scale and complexity but will typically take four to eight days. If the building is a semi or terraced house, the adjoining buildings will require support following demolition, adding to the cost.

If there is specialist work required, such as removing asbestos (often found in the form of cladding, roofing and rainwater goods) this can complicate issues. There are strict rules on the removal and handling of asbestos, and it can be best to get a report and quote from a specialist contractor.

Financing costs

These will include any application and valuation fees plus any inspection fees and interest payments. These costs will depend on how much you borrow and from whom you borrow. If you have your own money these costs will be nil. Most people don't so you will need to factor these in. You will get a better idea of these charges when you reach the stage of applying for a self-build mortgage. See chapter 3 for a further discussion of self-build mortgages on the market at the moment (2022).

Payment of professional fees

Once you have secured a plot you will almost certainly need an architect or a qualified designer to draw up the plans for your project. You will probably also need other professional help – for example a structural engineer and/or a quantity surveyor.

Sometimes self-builders appoint a project manager to run the entire project for them. They would appoint the rest of the professional advisers and organise a contractor for you, then manage their work. The Chartered Institute of Building may be able to point you towards local contacts. But there's really no substitute for personal recommendation.

Fees for planning approval and building regulation

These fees are standard and can be obtained from the local authority involved. They will be partly dependant on the value of the work.

Building materials and construction (labour) costs

In chapter 9, we will be looking more specifically at the costs of building and discussing construction costs in more detail. Here we look broadly at the costs which will allow you to arrive at a figure for borrowing.

The cost of building materials can vary but basically will depend on the quality of the materials selected and from whom they are sourced. If you are working to a strict budget then you will want to keep costs down. The following are main points to consider:

> If you employ a main contractor to undertake your self-build on a fixed price, then you will not be involved with materials selection unless this forms a part of the contract, i.e. choosing kitchens and bathrooms and finishes.

> If you are more involved with the project management, then you may well spend a lot of time phoning suppliers and going back and forward to builders merchants. Start by taking your plans around several local merchants-many have self-build specialists-and ask them to prepare a quotation based on your plans. Other things that you can do are:

> Open accounts with local Builders merchants. It makes negotiations much easier, and it helps with the record keeping..

> Open an account with a tool hire shop.

> Use the internet. It's especially good in tracking down fixtures and fittings like bathrooms, light fittings, plumbing gear, ironmongery.

Bulk materials (bricks, blocks, timber, plasterboard) are much cheaper ordered in quantity. Make sure that you know what quantity to order, and that you have somewhere on site to store them. Bricks and blocks may well require a digger or rough-terrain fork lift to unload and to place in the correct position.

Many builders merchants have very useful cost calculators which will give you a broad idea of what you will pay, depending on where you live. The calculations are usually always based on the Home Building and Renovations cost calculator. If you do your calculations using these sites you will arrive at a ballpark figure which will help you in putting together your budget for construction. Remember, the costs are for building only and don't include associated fees. The examples below are for costs in 2022.

Example build costs taken from a typical Build Costs Calculator

Using a proposed 120sm property in two areas:

1. Greater London
There are a number of options on the site which will provide different out turn costs. I have chosen two:

1. Using a single main contractors
£1454 per *sm* or £174,000 build cost

2. Self-managed using subcontractors
£1236 per *sm or £148,320*

2. Midlands

1. Using single main contractor

£1108.80 *sm or £133056*

2. Self managed using subcontractors

£943.20 *sm* of £113,184

There are several other options provided for in the build calculator which will give indicative costs. However, looking at the above, it can be seen that costs in Greater London are significantly higher than in the Midlands and in both examples managing the project yourself is far cheaper than using a single main contractor to undertake the whole build. Which option you choose will depend on your willingness to get involved personally, how much control you want, how much time you have and, most importantly, your budget. The web offers some very good advice and guidance on self-build costs.

Value added tax (VAT)

One area that you need to factor into your overall cost calculation is that of VAT. VAT can be reclaimed on self-build, within limits defined by HMRC. Below is a brief description of the way VAT works when self-building a home.

Reclaiming VAT

The **DIY Housebuilders Scheme** (Notice VAT431NB for new builds and Notice VAT431C for conversions) allows you to reclaim from HMRC some of the VAT that you have paid out for your project. You can apply for a VAT refund on building materials and services if you're:

> ➢ building a new home
> ➢ converting a property into a home
> ➢ building a non-profit communal residence - eg a hospice
> ➢ building a property for a charity

The building work and materials have to qualify and you must apply to HM Revenue and Customs (HMRC) within 3 months of completing the work.

Eligibility The application form has guidance on what building projects and materials qualify for a VAT refund. The basics are listed below.

New homes

The home must:

> ➢ be separate and self-contained
> ➢ be for you or your family to live or holiday in
> ➢ not be for business purposes (you can use one room as a work from home office)

Builders working on new buildings should be zero-rated anyway and you won't pay any VAT on their services.

Conversions

The building being converted must usually be a non-residential building. Residential buildings qualify if they haven't been lived in for at least 10 years. You may claim a refund for builders' work on a conversion of non-residential building into a home. For other conversions builders can charge a reduced VAT rate.

Building materials

You may claim a VAT refund for building materials that are incorporated into the building and can't be removed without tools or damaging the building.

What doesn't qualify

You can't get a VAT refund for:

> building projects in the Channel Islands
> materials or services that don't have any VAT - for example, they were zero-rated or exempt
> professional or supervisory fees - for example, architects or surveyors
> hiring machinery or equipment
> buildings for business purposes

➤ buildings that can't be sold or used separately from another property because of a planning permission condition

➤ building materials that aren't permanently attached to or part of the building itself

➤ fitted furniture, some electrical and gas appliances and white goods, carpets or garden ornaments

How to claim

Fill in form 431NB to claim a VAT refund on a new build, or 431C to claim for a conversion. Send your claim form to:
HM Revenue and Customs (HMRC).
National DIY Team SO970
HM Revenue and Customs
Newcastle
NE98 1ZZ

You must claim within 3 months of the building work being completed. There can be only one claim so make sure everything is on your completion certificate.

What you need to include

You must include all the documents listed in the claim form as part of your application. If an invoice is in your business's name, you'll need proof that you've refunded the amount from your personal bank - for example, a bank statement. VAT invoices must be valid and show the correct rate of VAT or they won't be accepted. For invoices not in sterling, convert the amounts into sterling.

How long it takes

You'll usually get the refund within 30 working days of sending the claim.

Having worked out what it will cost you to build your home, including all associated costs you will now be in a position (unless you are fortunate enough to be able to self-finance your project) to apply for a mortgage. In chapter 3, we cover all aspects of applying for finance.

Now read an overview of the Main Points from Chapter 2 overleaf.

****

Overview of the Main Points from Chapter 2-Costs of Self-Build

- One of the main aims of self-building a house is to ensure that you end up with more money than you started with, in terms of the final value of the finished property.

- The amount of money spent on a plot will depend on a number of factors. This can range from no expenditure if you are using your own back garden, for example, to hundreds of thousands of pounds, depending on what you buy and where you buy it.

- A cost to watch out for is stamp duty when purchasing the land. This will depend on the cost and the threshold in force at the time

- There will also be professional fees and costs of finance and other associated costs to factor in to the final calculation.

- A Building Costs Calculator used by main builders merchants will give you an idea of final costs of the actual building.

Don't forget you can claim back VAT!

Chapter 3

Obtaining a Self-Build Mortgage

If you are not fortunate enough to be able to build a home using savings, or with the funds from the sale of your previous home, you will need to borrow some money to finance your build. This is where a self-build mortgage will come in handy.

In reality, you should be getting a mortgage in principle before you even start looking for a plot, working out costs and applying for planning permission. This is much the same as you would do before you start looking for a conventional house. It ensures that you don't waste your time if you are found to be ineligible for a mortgage

Generally speaking, the amount you can borrow to purchase the land will be 75 per cent of its current value, and for the build costs, again you can borrow around 75 per cent of the end value. If you already own the land or the property, you can borrow against the value of this, meaning you can borrow more of the build costs.

You may need to access savings to fund the build so you should make sure they are not locked up. Also, if you are selling your house you will have to check your mortgage is free of early repayment charges.

Make a detailed plan

A lender will want to see detailed plans for the property, a projection of costs and planning permission details. The whole application process can take up to five months on average. You will have to be clear on everything including the people and materials being used. Factors such as build type, construction method, materials, location, and schedule of costs will all impact which lenders will lend and how much.

Paul and Jayne Marshall

We were aware that the source of funding for self-build would be more limited than a traditional mortgage, and also that the interest rate would be slightly higher. We approached Derbyshire Building Society and went to them with the land and outline planning permission. This is a fundamental requirement of lenders who won't consider funding without outline permission, although you can get an in-principle agreement that will allow you to see what you could borrow.

They also wanted to see the quotation from the builder (build costs) covering labour and materials. In our case, for the build in question, we needed to borrow £40,000.

This was agreed and released in three stages over the course of the build. The interest rate for the borrowing was about 2% higher than the average residential mortgage.

What is the difference between an ordinary mortgage and a self-build mortgage?

A self-build mortgage can exist in all of the same varied forms as an ordinary mortgage, from repayment to interest only. The main thing that differentiates a self-build mortgage from any other mortgage is in its operation and the existence of stage payments. To be a true self-build mortgage, the payments must also take in the land purchase stages. There are, typically, 6 stages, with slight differences between mortgages for brick built and timber frame construction, as below.

Stage	Brick & Block	Timber Frame
1	Purchase of land	Purchase of land
2	Preliminary costs & foundations	Preliminary costs & foundations
3	Wall plate level	Timber frame kit erected
4	Wind & watertight	Wind & watertight
5	First fix & plastering	First fix & plastering
6	Second fix to completion	Second fix to completion

Payment of the stage payments

There are two types of self-build mortgage: the arrears type, where the stage payments are given as each stage of the build is reached, and the advance type, where the stage payments are released at the start of each stage of the build.

The arrears-type self -build mortgage

This is suitable for those who have a large cash injection of their own to put into the project. For those with lesser means, this kind of mortgage can lead to cash shortfalls during the build and delays whilst they wait for a fresh cash injection. Additionally, many lenders giving this kind of self-build mortgage will typically only release 75% of the costs during the build period and will keep a retention of the loan amount until the project is finished.

The advance stage payment mortgage

This means that the money is in the bank and, therefore, available at the point of need when labour and materials bills fall due-removing the need for short-term borrowing/bridging loans to cover the shortfall. Typically the advance payment mortgages give up to 90% of the value of the land, plus 90% of the building costs, up to a maximum of 90% of the eventual value of the home, subject, of course, to the borrower's income status.

As we have seen, money is typically released at six stages, starting with land purchase, through foundation work, wall plate level, watertight, and so on.

When are the fees due?

Self-build mortgages tend to have slightly higher charges than regular house mortgages. You might reasonably expect to pay £1-2,000 to set up a self-build mortgage by the time application fees, broker fees and completion fees have been taken into account. On the arrears stage payment mortgage, an interim valuation is required at each stage before funds are released. An advance stage mortgage does not require interim stage valuations but there are additional costs in getting the money in advance. This cost depends, to a degree, on the amount of additional cash-flow required and the amount that the self-builders are putting in.

For full details of all self-build mortgages types and an in-depth explanation of each go to:

https://www.buildstore.co.uk/mortgages-finance/project-types/self-build-mortgages.

This site lists all of the most recommended mortgage providers currently operating in the UK. Rates for self-build mortgages are typically higher than the normal residential mortgages. In addition to the above site you should search the web as there are many other self-build mortgages on offer.

Example mortgage costs.
£170,000 over 25 years
Initial rate 2.28% for 2 years (24 instalments of £738.79 pm).
Lender fee £555.
The overall cost for comparison 4.06% APR representative

Subsequent rate 4.8% variable for the remaining 23 years (276 instalments of £919.75 pm)

Total amount payable £272,136.57

The above are examples only and are current at 2022. For further information and up- to- date rates you should apply to the various lenders.

Now read an overview of the Main Points from Chapter 3 overleaf.

****

Overview of Main Points from Chapter 3-Obtaining a Self-Build Mortgage.

- Most people will require a self-build mortgage. You should be getting a mortgage in principle before you even start looking for a plot, working out costs and applying for planning permission.

- Generally speaking, the amount you can borrow to purchase the land will be 75 per cent of its current value, and for the build costs, again you can borrow around 75 per cent of the end value. If you already own the land or the property, you can borrow against the value of this, meaning you can borrow more of the build costs.

- A lender will want to see detailed plans for the property, a projection of costs and planning permission details. The whole application process can take up to five months on average. build mortgage   The   main thing that differentiates a self- from any other mortgage is in its operation and the existence of stage payments. To be a true self-build mortgage, the payments must also take in the land purchase stages.

PART 2

GETTING A PROJECT OFF THE GROUND

- The Design Process

- Planning Permission

- Building Control

Chapter 4

The Design Process

Designing Your Home

If you're a first time self-builder your head might be spinning. It will be hard to know how to start planning the design of your home. You will almost certainly need to consider the use of an architect or other design professional in assisting you. Bear in mind that these professionals have many years of experience. In addition, the final plans that the architect draws up will be necessary for obtaining planning permission and dealing with Building Control (see Chapters 5 and 6 for more about planning permission and building control).

The main point to remember is that design isn't just about how the house looks, it is about ensuring that you can move freely in the space.

Another very important point is that you should plan for the future, as the one thing that you do not want is to have to add further expensive extensions to accommodate needs that you did not foresee. Below, we look at the different routes to achieving the design that you want, from using architects to designing your own home.

The different routes to house design-how to go about it

1. Using professionals

Good design is more than how the house looks on the surface. It is about how your home performs, how the spaces interact and flow and also meet your needs. Architects, or similarly qualified professionals can all help you to get the result you're after – and if you're after a truly bespoke design, the money is invariably well spent.

Fees for architects and professionals

Hiring an architect for the full suite of services – from preliminary design to move-in–comes at a price. Most architects will work for either an hourly rate or for a fixed fee based on a percentage of the construction costs – or some combination of the two. Architects fees will be around 7% for new build and 11% for existing buildings (that is as a percentage of the total building cost including labour and materials – but not land).

Most architects are members of the Royal Institute of British Architects (R.I.B.A.) or Royal Institution of Architects in Scotland, but all working in the UK must be registered with the Architects' Registration Board. Qualified Architectural Technologists (MCIAT) also provide services in helping with design.

2. Do-It-Yourself design

Designing a home yourself from scratch can be a very risky business, because you may end up with something that looks and feels wrong, falls foul of Building Regulations and might be unachievable. However, if you feel that you want to go ahead with this option, it is now technically easier to plan your own home, with the support of a number of software design packages on the market such as:

www.nchsoftware.com

ww.roomsketcher.com

ww.the-self-build-guide.co.uk/house-design-software.

3. Bespoke designs

One of the best ways to ensure that your build cost is established as early as possible is to use a design and build package supplier where costs are an integral part of the design process. While many designers of contemporary homes may turn their noses up at the schemes package companies have to offer, many package companies are really pushing the boundaries of design and you might be surprised at what they have to offer.

In the main, this book deals with traditional bricks and mortar developments. However, if you choose to go down the timber frame route, companies, such as timber frame specialists Potton, www.potton.co.uk, can offer one of the easiest routes to a well designed home that meets your needs. These firms offer a selection of pattern-book designs, most of which have been built before and which are customisable to suit your needs.

At the point of engaging an architect, or other professional or embarking on putting together your own design, you should consider the key points below. The main message is carry out as much initial work yourself so you can provide a steer to whoever you choose.

Keep it Small-It's the number one rule of building budget homes and it applies just as much to contemporary homes as it does to traditional styles-keep your floor area down. However, as we mentioned above, use the space to ensure that you get what you want out of the building and don't have to engage in any expensive alterations at a later date.

Come up with a basic vision-When you have decided on the size of your building, picture what kind of home you want to live in (and the type of budget you have). The earliest stages of the design process are about how you define your needs.

Look through architecture publications-Reading home design magazines can be a good way to get the ideas flowing. You'll get a good overview of current design trends, along with styles that were fashionable in past decades. You should also visit show homes on new developments to get ideas on layout and design.

Take photographs of houses to study later. When you see a home that catches your eye, get your camera out and shoot it from as many angles as you can. Photos also make great reference materials when you get into the specifics of piecing together your own home. Also, it is useful to keep a design notebook recording details such as contractor information and so on .

Consider your individual needs. As we discussed above, it is important to get what you need at the design stage. This is where considerations like space, privacy, and specific building techniques will come into play. In order to determine the type of living space that's right for you, it may help to outline the number, ages, and relationships of the people who will be sharing it. A two-bedroom house might not be big enough if you're planning on starting a family, but a slightly larger home could provide the space you need without forcing you to give up on your preferred aesthetic.

Make a list of essential features. Under individualized headings for each of the main rooms, start naming the amenities you want. Be specific as the more information you can give to your architect or team of contractors, the closer your finished home will be to your original vision. As you move onto drafting a floor plan, you can begin sorting the items on your list based on what's practical, what's affordable, and what makes the most sense.

Sketch out a rough floor plan-Block out basic areas first— for example, you might include two bedrooms on one end of the lower level, with the master bed and adjoining bath across the hall. Leave some room in the center of your layout for an open family room or study, then fill in the other end with plots for the kitchen, laundry room, dining room, and other important spaces. To keep your floor plan from getting confusing, focus on completing one level at a time. Keep yours or your family's needs in mind and try to devise a configuration that promotes both comfort and convenience.

Paul and Jayne Marshall-Involvement at the Design stage

After receiving the working drawings, we initially drew up and presented our preferred designs to our architect who then came back with several alternatives, the chosen ones shown above. The idea was to make the best use of the plot at the outset, ensuring that the space afforded by the footprint was fully utilised so that no further work would be needed after completion.

We looked at new build show homes in order to get ideas for interiors, such as fitted wardrobes and also went

---

online to look at different lighting set ups. We looked at different kitchen designs and, finally, we decided that we would go for under-floor heating, which makes sure that the heat is more evenly distributed, each room having a thermostat, and also eliminates the need for radiators on the ground floor, although they would be needed on the first floor.

Overall, a lot of thought, time and effort went into the final design. We wanted to ensure that we maintained control of the process and have maximum input at the outset. This is particularly recommended for all first time self-builders who lack experience. Hindsight is a wonderful thing although it can be expensive when the house has finally been constructed!

See below for the initial (proposed) design creations for our current self-build. These show the ground and first floor of the dorma bungalow and also the whole layout. The designs came from the working drawings prepared by the architect, which will be shown in detail in the next chapter covering Planning Permission.

Fig 1 shows the first detailed drawing of the whole house with the ground and first (Dorma) floor laid out seperatly.

Fig 2 Shows the interior layout from the rear elevation.

Fig 3 Shows whole house

Fig 4 Shows overhead view of interior

Fig 5 Shows proposed patio layout

Fig 6 Shows internal ground floor layout-Kitchen diner

Fig 7 Shows diner/living area

Fig 8 Shows whole house

For a fuller view of the design drawings go to www.straightforwardco.co.uk (Straightforward Publishing website) and click on the book cover on the front page.

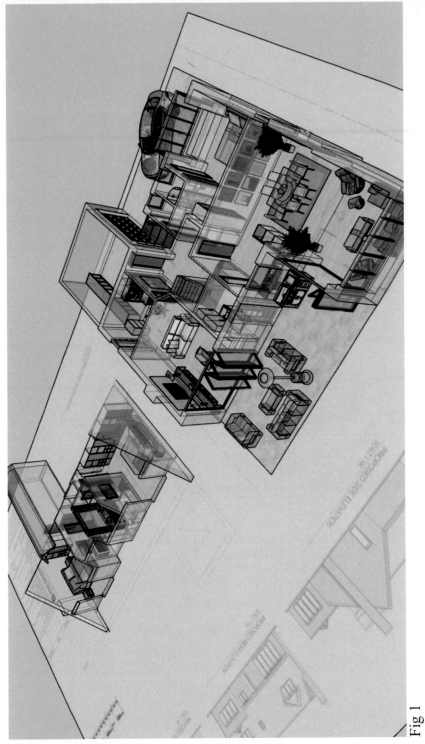

Fig 1

Fig 2

Fig 3

Fig 4

Fig 5

Fig 6

Fig 7

Fig 8

Working with our architect, and also with the input of our builder, we arrived at the design that we wanted plus, most importantly, bearing in mind that design is not all about how the property looks but also how it is used, we ensured that the interior space suited our needs entirely, both now and for the future, adapting to any changes in circumstances.

Now read an overview of the main points from Chapter 4 overleaf.

****

An overview of the Main Points From Chapter 4-The Design Process

- If you're a first time self-builder it will be hard to know how to start planning the design of your home. You will almost certainly need to consider the use of an architect or other design professional in assisting you.

- Most people employ architects to draw up initial plans to obtain planning permission and design features should be input at an early stage.

- Good design is more than how the house looks on the surface. It is about how your home performs, how the spaces interact and flow and also meet your needs

- Designing a home yourself from scratch can be a very risky business, because you may end up with something that looks and feels wrong, falls foul of Building Regulations and might be unachievable

- At the point of engaging an architect, or other professional carry out as much initial design work yourself so you can provide a steer to whoever you choose.

Chapter 5

Obtaining Planning Permission

This chapter covers the planning system in England. Although the planning systems are similar in each country in the UK, nevertheless there are differences in emphasis and procedure.

For more information about the planning system in Wales, Scotland and Northern Ireland you should visit:

> businesswales.gov.wales/running-business/premises-property-planning-building/planning (Wales)
> www.mygov.scot/planning-permission (Scotland)
> www.nidirect.gov.uk/information-and-services/repairs-planning-and-building-regulations/planning-system (Northern Ireland)

Planning permission generally

We have covered the preparation of design generally and the production of more detailed drawings that can be used for the purposes of obtaining planning permission and also for obtaining building regulations approval.

The construction of new buildings and extensive changes to existing buildings will usually require consent from the local planning authority in the form of planning permission. The planning system is designed to control inappropriate development. For details of your local authority you should go to www.planningportal.co.uk. where you can begin the process of obtaining planning permission. Anything that involves the creation of a new house, either by building from scratch or a subdivision, needs planning permission. Adding outbuildings or building extensions requires planning permission depending on the size of the project and the level of Permitted Development rights afforded to or still remaining on a property.

Paul and Jayne Marshall

The most important point when it comes to obtaining planning permission, whether outline or full, is that you need to use an architect. As we saw in the previous chapter, they will let you know what they would charge for taking on the project and steer you through the design process, along with making the planning application. They will also deal with building control issues. In addition, they can arrange for a topographical survey and also deal with the local utilities, such as water authorities who need to provide clearance before you can build.

There are many issues when seeking to obtain planning

permission to construct a property. The most important ones, in our case, were those of access, due to the proximity of the main road, and traffic coming out onto that road and also privacy issues, light issues and so on.

We found that the delegated planning powers given to staff in the planning department were enough to get through most hurdles but, with one of our projects we encountered complex issues and the application eventually went to full committee.

At the end of this chapter we have included the drawings put together by our architect, Paul Gaughan, Building Consultants, which enabled us to get the proposed construction through planning and also which provided working drawings for our builder and assisted in the design process. In the index is a copy of the building specification and full planning permission relating to the project.

Permitted Development Rights

The policy of Permitted Development was introduced at the very beginning of the planning system – in the Town and Planning Act 1948, on 1st July 1948 – and allows for minor improvements, such as converting a loft or modest extensions to your home, to be undertaken without clogging up the planning system. Scotland, Wales and Northern Ireland each benefit from their own version of these rules. You should visit the websites outlined above.

The level of work that can be carried out under Permitted Development depends on a variety of factors including location (Areas of Natural Beauty and Conservation Areas have different rules), and the extent of work already carried out on a property

Outline and Full planning permission

Outline planning permission grants, in principle, the construction of a dwelling, subject to certain design conditions based on size and shape. If your plot comes with Outline permission, you will need to examine the approval document, which will give you a good idea of the type of house you could end up building. Full approval is what it says and is likely on a design that follows these guidelines. However, it is not absolutely set in stone as other design schemes could be approved.

Costs of applying for planning permission

The fee for submitting a planning application varies depending on the nature of the development. The cost is currently £385 for a full application for a new single dwelling in England, but this fee is different in Scotland, Wales and Northern Ireland. For home improvers, an application in England for an extension currently costs £172, whereas in Wales the cost of a typical householder application is currently £166.

As well as fees for pre-application advice, further small sums are payable for the discharge of 'planning conditions' which must be met before development begins.

Planning application requirements

Each site has different requirements but, in general, an application must include:

- five copies of application forms,
- the signed ownership certificate,
- a site plan, block plan and elevations of both the existing and proposed sites,
- a Design and Access Statement
- the correct fee.

Design and Access Statements

Design and Access Statements have to accompany all planning applications. Statements are used to justify a proposal's design concept and the access to it. The level of detail depends on the scale of the project and its sensitivity. Most authorities will have guidance notes available to help you. Unless you ensure you have included one in your submission, planning authorities can refuse to register your planning application.

Planning Conditions

Planning permission can be subject to planning conditions which need to be discharged/agreed within a given time.

Planning conditions are extremely important and failure to comply can result in what is called a breach of condition notice, to which there is no right of appeal-not to mention it could be enforced through the courts by prosecution. Planning Conditions can be as simple as requiring that building materials must match existing ones in the area or on the building if a refurbishment.,

Planning decisions

Local authority planning committees, and also planning officers using delegated powers, meet to decide planning applications. They will base their decisions on what are known as 'material considerations', which can include:

> ➤ Overlooking/loss of privacy
> ➤ Loss of light or overshadowing
> ➤ Parking
> ➤ Highway safety
> ➤ Traffic
> ➤ Noise
> ➤ Impact on listed building and Conservation Area
> ➤ Layout and density of building
> ➤ Design, appearance and materials
> ➤ Government policy
> ➤ Disabled access
> ➤ Proposals in the development plan
> ➤ Previous planning decisions
> ➤ Nature conservation

There may be other factors involved depending on the scheme. While neighbours are consulted and invited to comment, together with parish councils (in England and Wales), only objections based on material considerations are taken into account. If the neighbours do not object and the officers recommend approval, they will usually grant planning permission for a householder application using delegated powers.

If there are objections or the application is called into a committee by one of the local councilors, then the decision will be made by a majority vote by the local planning committee. At the planning meeting, you or your agent will be given an opportunity to address the planning committee, but this time is limited to a maximum of three minutes.

Timeframe for planning approval

Once your application has been submitted, the planning department will check that all of the information it requires has been received together with the correct fee. The time period for planning decisions is 10 to 12 weeks following registration, The majority of straightforward applications will be dealt with within this timeframe.

During the approval process, a sign is posted outside the address relating to the proposed development and any neighbours likely to be affected are written to and invited to view the plans and to comment. This is the public

99

consultation process and will take three to eight weeks. The authority will make statutory consultations to the local Highways department, and where necessary the Environment Agency as well as others.

Planning application refused

It is the case that in England around 75 per cent of applications are granted. It is not the end of the world if your application is refused. there will usually be a good and clear reason and the refusal will be based on something that you can rectify and resubmit the application..

If your application is refused, you can also make an appeal to the planning inspectorate — around 40 per cent of householder applications that are refused are later granted at appeal.

Planning application approved and commencement of work

Planning permission is typically granted for three years — meaning you must begin work in that time or face reapplying. Nearly all people will start work straight away, if intending to build to live in the property.

You can make minor alterations to plans that have been approved by applying for a non-material amendment. However, major alterations could involve a further application for Full planning permission. You will need to discuss your plans with your Local Planning Authority first.

Carrying out works without planning permission

Highly inadvisable to do this. Although not illegal to develop land without planning permission, it is not lawful and, consequently, if you have failed to get consent for your project, then the local planning authority can take action to have the work altered or demolished. In this instance, you can make a retrospective planning application and if this is refused you can appeal the decision. If you lose, it can prove very costly.

There is a legal loophole: if no enforcement action is taken within four years of completion, the development becomes immune from enforcement action (10 years for a change of use). The development then becomes lawful.

Altering a listed building without prior permission is, however, a criminal offence, and in extreme cases it can lead to prosecution and unlimited fines — and even imprisonment. .

Extending planning permission

If you were granted planning on or before 1st October 2009, and it has not already run out, you can apply to extend it. The form is available on The Planning Portal and is subject to a fee of £50.

See overleaf for detailed drawings produced for Paul and Jayne Marshall's self-build by their architect. Following on from the drawings, see a copy of the final planning consent in Appendix 1. For a fuller view of the drawings and planning consent go to www.straightforwardco.co.uk and click on the book cover on the front page.

Fig 1-Topographical survey

Fig 2-Proposed drainage site block plan

Fig-3-Proposed foundation layout

Fig 4-Prposed ground floor layout

Fig 5-Prposed first floor layout

Fig 6-Interior sectional drawing

**

PAUL GAUGHAN
BUILDING CONSULTANTS

TOPOGRAPHICAL SURVEY
SCALE 1:200

DBP/JM/14/049/01

PROPOSED FENCING DETAIL
SCALE 1:50

PROPOSED DRAINAGE SITE BLOCK PLAN
SCALE 1:200

PROPOSED LANDSCAPING SITE BLOCK PLAN
SCALE 1:200

DRAINAGE SCHEDULE

LANDSCAPING SPECIFICATION

PAUL GAUGHAN
BUILDING CONSULTANTS

DRAWING NUMBER
DB/P.JM/14/04.9/10

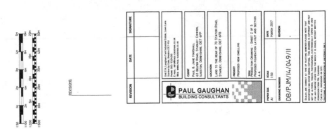

PAUL GAUGHAN
BUILDING CONSULTANTS

DRAWING NUMBER
DB/PJM/14/049/11

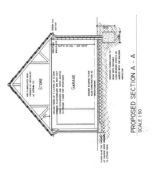

PROPOSED SECTION A - A
SCALE 1:50

STORE

GARAGE

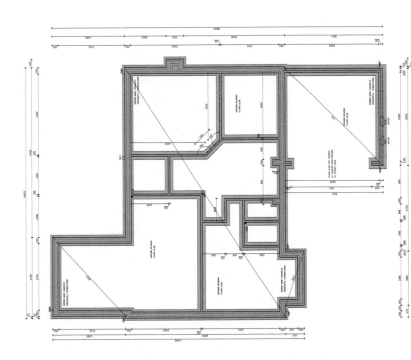

PROPOSED FOUNDATION LAYOUT
SCALE 1:50

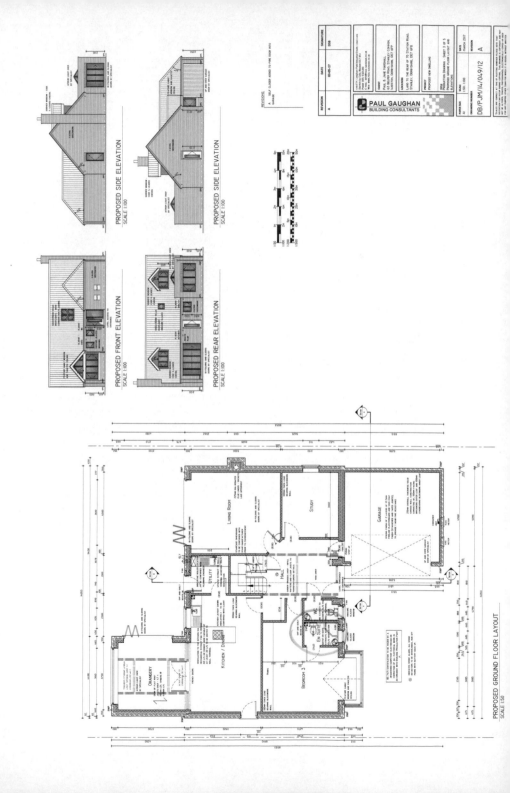

PROPOSED SIDE ELEVATION
SCALE 1:100

PROPOSED SIDE ELEVATION
SCALE 1:100

PROPOSED FRONT ELEVATION
SCALE 1:100

PROPOSED REAR ELEVATION
SCALE 1:100

PROPOSED GROUND FLOOR LAYOUT
SCALE 1:50

PAUL GAUGHAN
BUILDING CONSULTANTS

DRAWING NUMBER
DB/P.JM/14/04.9/12

REVISION
A

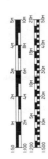

REVISIONS

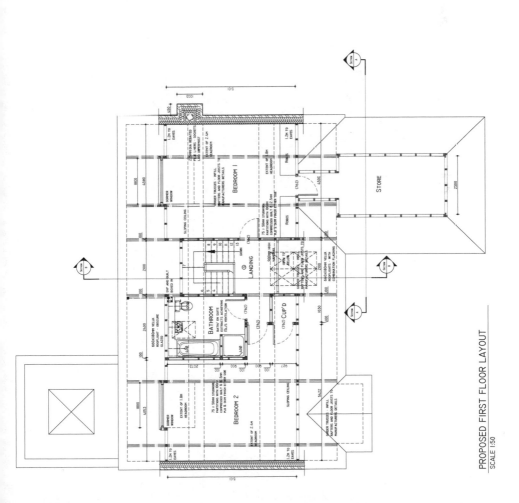

PROPOSED FIRST FLOOR LAYOUT
SCALE 1:50

REVISION	DATE	SIGNATURE

PAUL GAUGHAN
BUILDING CONSULTANTS

Unit 14, Langham Park Industrial Estate, Lows Lane
Stanton by Dale, Ilkeston DE7 4RJ
Phone: 015 9324240
E-MAIL: INFO@PAUL-GAUGHAN.CO.UK
WEB: WWW.PAUL-GAUGHAN.CO.UK

CLIENT
PAUL & JANE MARSHALL
63 BELPER ROAD, STANLEY COMMON,
ILKESTON, DERBYSHIRE, DE7 6FP

LOCATION
LANE TO THE REAR OF 70 STATION ROAD,
STANLEY, DERBYSHIRE, DE7 6FB

PROJECT
PROPOSED NEW DWELLING

TITLE
PRODUCTION DRAWING - SHEET 4 OF 5
PROPOSED FIRST FLOOR LAYOUT

PAPER SIZE	SCALE	DATE
A1	1:50	MARCH 2007

DRAWING NUMBER | REVISION
DB/PJM/14/04.9/13 | .

SCALES ARE CORRECT AT TIME OF PLOTTING HOWEVER PLEASE NOTE THAT
DISTORTION CAN OCCUR DURING COPYING. THIS DRAWING IS COPYRIGHT AND MAY
NOT BE ALTERED, TRACED, PHOTOGRAPHED, ELECTRONICALLY COPIED OR USED
FOR ANY PURPOSE OTHER THAN FOR WHICH IT IS ISSUED, WITHOUT WRITTEN
PERMISSION FROM MR. GAUGHAN.
CHECK ALL DIMENSIONS ON SITE.
HATCHING IS A REPRESENTATION OF MATERIALS ONLY.

BEDROOM 1

BEDROOM 2

BATHROOM

LANDING

CUP'D

STORE

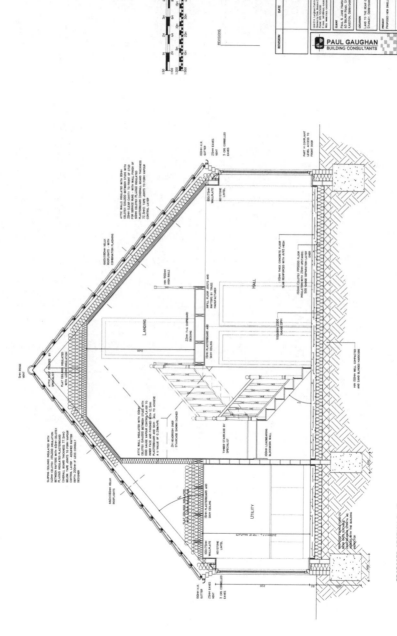

PROPOSED SECTION B - B
SCALE 1:20

PAUL GAUGHAN
BUILDING CONSULTANTS

CLIENT
PAUL & JANE MARSHALL

LOCATION
LAND TO THE REAR OF 70 STATION ROAD,
STANLEY, DERBYSHIRE, DE7 6FB

PROJECT
PROPOSED NEW DWELLING

TITLE
PRODUCTION DRAWING - SHEET 5 OF 5
PROPOSED SECTION B - B

DRAWING NUMBER
DB/P-JM/14/04.9/14

Summary of the planning process

PLANNING APPLICATION SUBMITTED

COUNCIL CHECKS WHETHER THE APPLICATION IS VALID AND REQUESTS ANY MISSING PAPERWORK

COUNCIL ACKNOWLEDGES THE APPLICATION IS VALID

COUNCIL PUBLICISES APPLICATION IN ACCORDANCE WITH ITS POLICY AND WRITES TO ANY STATUTORY CONSULTEES (E.G. ENVIRONMENT AGENCY)

COUNCIL PLANNING OFFICERS WRITE A REPORT WITH RECOMMENDATIONS

COUNCIL'S DELEGATION SCHEME SETS OUT WHO SHOULD MAKE THE DECISION

COUNCIL'S CHIEF PLANNING OFFICER COMMITTEE

COUNCIL'S PLANNING

DECISION

REFUSE_____

GRANT WITH

CONDITIONS

So-assuming that you have found a plot of land, obtained the funding to buy it, and have at least outline planning permission we now need to look at the process of dealing with local authority building control.

Now read an overview of the Main Points from Chapter 5 overleaf.

****

An overview of the Main Points from Chapter 5-The Planning Process

- The construction of new buildings and extensive changes to existing buildings will usually require consent from the local planning authority in the form of planning permission.
- The policy of Permitted Development Rights allows for minor improvements, such as converting a loft or modest extensions to your home, to be undertaken without clogging up the planning system
- Outline planning permission grants, in principle, the construction of a dwelling, subject to certain design conditions based on size and shape. Full approval is likely on a design that follows these guidelines. However, it is not absolutely set in stone as other design schemes could be approved.
- Design and Access Statements have to accompany all planning applications. Statements are used to justify a proposal's design concept and the access to it. The level of detail depends on the scale of the project and its sensitivity
- The time period for planning decisions is 10 to 12 weeks following registration.

Chapter 6

The Role of Local Authority Building Control

Building control generally

After all the hard work and expense of producing drawings, producing final designs and applying for planning permission, comes the process of working with local authority building control.

Each local authority has a building control department which acts as an independent assessor of your project and proposals. They ensure that the plans of your new home, together with the materials and workmanship which go into the construction, comply with the legal requirements laid down by Parliament and which are expressed as Building Regulations. Virtually all self-build projects will be dealt with by the local authority building control office.

Building Regulations apply to new homes without exceptions, specifying a minimum standard of performance which must be achieved. Self-builders may therefore consider that they can improve on the minimum standards required and this can be done in a manner of means which will enhance the value and efficiency of the house – for example larger floor joists, improved levels of heating and ventilation.

The main purpose of Building Regulations is to secure health and safety of the people in the buildings but they also deal with energy conservation, providing access and other matters. Detailed plans must be prepared showing all constructional details and should be submitted well in advance of commencement of work on site. There is a fee involved and the local authority must pass or reject your plans within a certain time limit or they may, with your agreement, add conditions to their approval. They must, if satisfied with your proposals, issue an approval notice within five weeks, or, if you agree to an extension, no longer than two months.

If the plans are rejected then there is a procedure for a review. Building Control will also act in regular inspections of the work as it goes along and you should treat your Building Control officer as an additional and impartial pair of eyes to inspect work in progress. Once the local authority are satisfied that the works have been carried out and meet the Building Regulations a Certificate of Completion can be issued. This is a valuable document which should be retained safely, probably with your title deeds.

If you are carrying out building work personally, it is very important that you understand how the building regulatory system and material applies to your situation as you are responsible for making sure that the work complies with the building regulations.

More about procedures

If you are employing a builder, the responsibility will usually be theirs - but you should confirm this at the very beginning. You should also bear in mind that if you are the owner of the building, it is ultimately you who may be served with an enforcement notice if the work does not comply with the regulations.

Some kinds of building projects are exempt from the regulations, however generally if you are planning to carry out 'building work' as defined in regulation 3 of the building regulations, then it must comply with the building regulations. This means that the regulations will probably apply if you want to:

- Put up a new building
- Extend or alter an existing one
- Provide services and/or fittings in a building such as washing and sanitary facilities, hot water cylinders, foul water and rainwater drainage, replacement windows, and fuel burning appliances of any type.

The works themselves must meet the relevant technical requirements in the building regulations and they must not make other fabric, services and fittings less compliant than they were before-or dangerous. For example, the provision of replacement double-glazing must not make compliance worse in relation to means of escape, air supply for

115

combustion appliances and their flues and ventilation for health.

They may also apply to certain changes of use of an existing building. This is because the change of use may result in the building as a whole no longer complying with the requirements which will apply to its new type of use, and so having to be upgraded to meet additional requirements specified in the regulations for which building work may also be required.

In summary, the following types of project amount to 'building work':

- The erection or extension of a building
- The installation or extension of a service or fitting which is controlled under the regulations
- An alteration project involving work which will temporarily or permanently affect the ongoing compliance of the building, service or fitting with the requirements relating to structure, fire, or access to and use of buildings
- The insertion of insulation into a cavity wall
- The underpinning of the foundations of a building
- Work affecting the thermal elements, energy status or energy performance of a building.

Now read an overview of the main points from Chapter 6 overleaf.

****

An overview of the Main Points from Chapter 6-Building Control

- Each local authority has a building control department which acts as an independent assessor of your project and proposals. They ensure that the plans of your new home, together with the materials and workmanship which go into the construction, comply with the legal requirements laid down by Parliament and which are expressed as Building Regulations. Virtually all self-build projects will be dealt with by the local authority building control office. Building Regulations apply to new homes without exceptions, specifying a minimum standard of performance which must be achieved

- The main purpose of Building Regulations is to secure health and safety of the people in the buildings but they also deal with energy conservation, providing access and other matters.

- The local authority must, if satisfied with your proposals, issue an approval notice within five weeks, or, if you agree to an extension, no longer than two months.

- Once the local authority is satisfied that the works have been carried out and meet the Building Regulations a Certificate of Completion can be issued.

PART 3

UNDERSTANDING THE CONSTRUCTION PROCESS

■ General Overview and Construction Contracts

■ Trades-People Involved

■ Estimating Construction Costs

■ The Stages of Construction

Chapter 7

An Overview of the Construction Process

Construction management generally

Construction management entails setting up a program of construction, organising a supply of materials to site and employing and supervising the workforce. In addition, you will need to put in place the necessary insurances while on site and also arrange for stage payments. This all sounds complicated, and it certainly can be unless you have a grip on matters. If you have chosen a bespoke option then all of this will be carried out by the main contractor. If you have put everything in the hands of an architect then the process will be managed for you. However, if you have decided to go ahead on your own then you need to understand the processes.

The management of the construction process deals with the most complex organisational and financial elements of self-build. The stages leading up to getting on site are also complex but nothing compared to the building of your house. One thing is for sure, how far you go in the involvement of construction will depend on how much time you have got, how much experience you have, how much money you have and the type of property you are constructing. In all likelihood you will be paying someone to get involved for you, but it is still very useful to know the ins and out's of what is happening.

We will start by looking at the contract that you will have with your builder at the start of the process.

The contract with your builder

As we have seen, construction projects are usually complex affairs, with a number of different trades working on the site. Following on from this, there are risks that can arise. In this situation, contracts are needed because they provide a framework for dealings between the participants helping to identify how risks are allocated and what to do if things go wrong.

In short, contracts are important because:

- They record the commercial terms agreed by the parties
- They provide certainty during the construction process
- They provide a framework for dealing with issues that might arise during the life of the contract
- They provide a method of resolving disputes

In addition, you will need a clear contract to deal with a whole range of issues, such as scope of work, price and payment terms, variations, Insurance, contract duration and completion date, post-completion liability, insurance, termination and dispute resolution.

Anyone taking on work that involves constructing or enlarging a dwelling has a legal duty to "see that the work which he takes on is done in a workmanlike or... professional manner with proper materials so that ...the dwelling will be fit for habitation when completed". This duty is set out in the Defective Premises Act 1972 and claims can be brought for six years after completion.

The law applies to architects and other professional consultants as well as builders and other tradesmen involved in building a home. It also applies to extensions and conversions, not just new properties.

The JCT (Joint Contract Tribunal) www.jctltd.co.uk has been producing standard forms of contract for the construction industry since 1931. They offer a downloadable contract designed for self-builders, renovators and home improvers, who are directly employing a builder or consultant such as an architect.

The various areas you will find covered in a typical contract are:

- Procurement Structure-(How you intend to organize the build)
- Scope of Work or Services

The contract should include a schedule which sets out the scope of the work– probably by reference to a detailed specification document and a set of numbered drawings.

- Contract Price

There are different ways of pricing a contract. A fixed price or lump sum for example. It can be useful to have a schedule that breaks the price into a number of different elements – foundations, main structure, roof etc (see chapter 9). Then if there is a change, you can have more control. Other pricing options include:

- Time & Materials, where a builder perhaps gives an estimate but charges for the time spent at pre-agreed rates plus the cost of materials.
- Cost plus Fee where a builder charges the actual cost of labour and materials plus a fee to cover his management expenses and profit.
- Payment Terms

There are two main ways of arranging payment during the contract:

- regular weekly or monthly invoices from the builder based on work done, or
- a fixed payment schedule with specified amounts payable at particular stages of the work

When agreeing the payment terms, try to ensure that the money paid out matches the work done on site. This limits your risk if the builder gets into financial difficulties.

Advance payments to cover the cost of materials are sometimes agreed

- Work on Site

Especially when the project involves work to your home, it is sensible to have a clause that sets out the rules of conduct such as working hours, compliance with health and safety regulations, co-operation with other contractors and the liability of the contractor for any damage caused to property of the client or of the neighbours. Cleaning up after work and safe storage of plant and equipment can also be covered in the contract.

- Changes

Not every client knows all the details of what he/she wants when the project starts and even if they do they may change their mind. The contract should cater for this with a change or variations clause. This will set out a procedure for dealing with change.

- Delays & Time Extensions

With all construction projects, unforeseen events can occur that change the price and programme. For example, unforeseen ground conditions can impact on the project

125

cost as can other events such as severe weather or delay caused by utility companies.

The contract should provide that the contractor has to notify the client once he knows of something beyond his control that could delay completion and the client has to decide whether to give an extension of time and revise the completion date. Always be sure to revise the completion date if there is some delay or change that prevents the contractor from completing on time. If you do not do this, the contractor is no longer tied to a firm date and his contractual obligation to complete the work is watered down.

- Completion & Defects Liability

It is usual to have a start date and a completion date in a building contract. When the work is finished, it will be inspected by the client or architect and the completion date will be confirmed. It is then usual for a contract to have a defects period of six or twelve months during which any problem that arises has to be put right by the contractor. However, the potential liability of the contractor will not stop at the end of the defects period unless the contract says so. Under the general law, a contractor is liable for any latent defect that appears for up to six years from completion or twelve years if the contract is executed as a deed.

Also anyone taking on work that involves building or enlarging a dwelling has a legal duty to "see that the work which he takes on is done in a workmanlike or... professional manner with proper materials so that ...the dwelling will be fit for habitation when completed". This duty is set out in the Defective Premises Act 1972.

- Liquidated Damages

Most large building projects contain provision for the contractor to pay liquidated damages (LDs) if they fail to finish the work by the completion date. Liquidated damages are a pre-agreed amount – e.g. £500 a day or £500 a week - and should represent an estimate of the likely cost of delay to the client such as the cost of renting another property. LDs are not often used in small projects but they could be appropriate for a main contractor building a £300,000 house.

- Insurance & Liability

Normally a contractor for the main building works will have Contractor's All-Risk Insurance, Public Liability Insurance and Employers' Liability Insurance. There will be a minimum level of cover specified for these insurances.

An architect, other professional consultant (and a contractor) who undertakes design, should have professional indemnity insurance. Architects and other professionals may insist on having a limit on liability in their contract with you. This is reasonable, provided that the limit is an adequate one, which will usually be the case as it will be backed by insurance. Builders too may want a limit, which, again, can be accepted if the limit is a fair one for you, the client.

- Ownership & Risk

Partly to protect against the risk of a contractor bankruptcy, a clause dealing with ownership of materials is recommended. This will usually say that ownership of materials supplied by the contractor will pass to you, the client, when the materials arrive on site or when they are paid for, if this is earlier. Risk of loss or damage will, however, remain with the contractor until he hands over the completed work.

- Termination & Suspension

It is always advisable to have a termination clause which states the grounds on which a party can bring the contract to an end and any relevant consequences. It is usual to include breach of contract and insolvency as reasons for terminating. If the contractor is in breach, the contract wording should say that the client is allowed to take over all plant and equipment on the site and does not have to pay the contractor until the work done by a replacement contractor has been assessed. The extra cost and loss caused by the failed contractor will then be taken into account.

Sometimes the client has the right to terminate at any time by giving notice but in that case, you would usually have to pay reasonable demobilisation costs to the contractor and possibly compensation for loss of profit. If problems arise, a client occasionally wants the right to suspend the contract - e.g. if planning consent is delayed. The contract can contain a clause that allows you to suspend the contract but if that continues for more than a specified period, say 3 months, it is usual to state that either party can then terminate the contract.

- Longer-Term Risks

When building a new house you need to think about the long term consequences. Even if you stay in the house for several years, if some defect appears after 5 years you might have a claim against your architect or builder but it could be expensive to pursue and there is always the possibility of the builder going bust in the meantime.

Also, if you sell, the buyer will want to be sure that your building work has all the necessary consents from the local authority. He/she may also want to have some protection in case defective work appears after he has moved in. Under the general law he/she would not find it easy to sue your builder or architect. Taking out a 10-year structural warranty policy is a good way to overcome this problem.

- Disputes

Building disputes can be time consuming and expensive, and there are interim procedures cheaper than the courts including adjudication and mediation.

Adjudication is a short-term procedure where an adjudicator has to make a decision within twenty-eight days of a dispute being referred to him. Mediation involves an independent mediator who facilitates a negotiated solution. These methods can keep the costs down and achieve a quicker and cheaper solution than going through the courts.

Now read an overview of the Main Points from Chapter 7 overleaf.

****

An overview of the Main Points from Chapter 7-The Construction management process

- Construction projects are usually complex affairs, with a number of different trades working on the site. Following on from this, there are risks which can arise. In this situation, contracts are needed because they provide a framework for dealings between the participants helping to identify how risks are allocated and what to do if things go wrong.

- Contracts are important because:
  o They record the commercial terms agreed by the parties
  o They provide certainty during the construction process
  o They provide a framework for dealing with issues that might arise during the life of the contract
  o They provide a method of resolving disputes

- In addition, you will need a clear contract to deal with a whole range of issues, such as scope of work, price and payment terms, variations, insurance, contract duration and completion date, post-completion liability, insurance, termination and dispute resolution.

Chapter 8

Main Trades People Involved in Construction

Builders and Subcontractors

Having ensured that you that you understand the general process of construction and the importance of the contract, you then need to establish a good relationship with a builder. Although there are good reputable builders out there, there are also cowboys who can ruin your life! If you intend to manage the project yourself, you also need to establish relationships with sub-contractors.

In most cases however, you will choose a builder who will act as a main contractor, who will manage the whole process either on a labour only basis, leaving you to purchase the materials, or who will do the lot for you. In any event, it is important to understand the roles of the various trades involved in the construction of a house.

A **ground worker** is a term for a subcontractor who is employed to prepare a home construction site for the shallow foundation of a new home. Typically, the ground worker clears the site, lays a foundation, installs drainage and other pipework, and may build roads if necessary. The ground workers are usually the first and last trades performed on site. They will start the job putting in levels, digging the ground, excavating and concreting the foundations, and building the foundations until the floor is on and the work is up to the damp proof course (dpc). They are also the team responsible for the drainage; connecting to the existing pipes and installing the new pipes. At the end of a home build, ground workers return to put in driveways and footpaths.

Bricklayers build and maintain a range of structures including internal and external walls on new houses and commercial projects, chimney stacks, tunnel linings and decorative work such as archways and garden walls. They also deal with the repair and refurbishment of existing brickwork or masonry. They use a variety of hand and power tools and various materials such as bricks, blocks, lintels, stone and mortar, depending on the job.

A typical new-build housing job will include the following tasks:

- measuring the area and setting out first courses and damp course, following architects' or designers' plans
- working from the corners, building up the courses using bricks and mortar (for efficiency, bricklaying teams or "gangs" often employ a labourer to keep them constantly supplied with bricks and mortar)

- using a variety of hammers and other tools to shape and trim bricks, trowels to apply mortar, and spirit-levels and plumb-lines to check courses are correctly aligned.
- As the building goes up, bricklayers or scaffolders will raise platforms to reach the higher storeys; joiners usually follow closely behind fitting frames for doors and windows as designated by the design blueprints. Depending on the size of the project, gangs of bricklayers may work on different sections at the same time. Some bricklayers may also specialise in stonemasonry work.

Carpenters construct and repair building frameworks and structures—such as stairways, doorframes, partitions, and rafters—made from wood and other materials. They also may install kitchen cabinets, siding, and drywall. Carpenters typically do the following:

- Follow blueprints and building plans to meet the needs of clients
- Install structures and fixtures, such as windows and moulding
- Measure, cut, or shape wood, plastic, and other materials
- Construct building frameworks, including walls, floors, and doorframes

- Help erect, level, and install building framework with the aid of rigging hardware and cranes
- Inspect and replace damaged framework or other structures and fixtures
- Instruct and direct labourers and other construction helpers

Carpentry is one of the most versatile construction occupations, with workers usually doing many different tasks. For example, some carpenters insulate office buildings; others install drywall or kitchen cabinets in homes. Those who help construct tall buildings or bridges often install the wooden concrete forms for cement footings or pillars. Some carpenters erect shoring and scaffolding for buildings.

Plasterers are building professionals who apply plaster to walls and ceilings to create a finished look to an interior space. They may work with new construction or manage repairs to existing buildings. In addition to the creation of basic coverings on walls and ceilings, plastering may also involve the creation of architectural details that enhance the overall look of the space.

For basic plastering jobs, the plasterer will begin with either a solid surface or wire lathing that is attached to the surface. For many projects, an even coat of gypsum is applied to the facing of the wall or ceiling to create a

136

surface that the plaster will adhere to with no problems. The use of gypsum is particularly helpful when the plaster professional is working with a concrete wall.

With the gypsum in place, the builder will apply what is typically known as the finish coat. This coat is usually plaster that is formulated with a lime base.

Plumbing. In a residential setting, plumbers both install new plumbing fittings and repair or replace existing fixtures. This includes plumbing related to water coming into the house, waste exiting the house and gas lines needed to power appliances, such as hot water heaters. Plumbers also install, repair or maintain the plumbing to toilets, shower and bath systems, kitchen fixtures such as faucets and dishwashers, water heaters and septic tanks. Residential plumbers also lay new pipes in homes under construction, which can involve cutting pipe, sawing wood and welding.

Electricians working in new construction typically work from blueprints that an architect or builder has produced. These blueprints tell the electrician where outlets, switches, circuit breakers and lighting fixtures need to go. The electrician then determines the proper way to run the wiring needed. Electricians install conduits to hold wiring when needed, run the wires and connect them. Electricians who are still serving their apprenticeships cannot fill these roles without direct supervision.

Painters and decorators are tradesman responsible for the painting and decorating of buildings, and are also known as a decorator or house painter. The purpose of painting is to improve the aesthetic of a building and to protect it from damage by water, rust, corrosion, insects and mould.

Wall and floor tilers affix ceramic, slate, marble and glass tiles to walls and floors, using glues, grout and cement. They frequently cut tiles in order to fill small edges or make particular patterns. Tiles provide both a decorative and protective function, especially in spaces that experience ongoing wet or damp conditions such as kitchens or bathrooms.

Roofers are professionals who specializes in roof construction. Roofers monitor the entire process of roofing in residential as well as commercial construction. They analyse the construction plans and make sure that the roofing is done in strict accordance with the design. Roofers also determine the materials, substrates and supportive accessories to be used for roof installations. Even the specifications of the beams, trusses and rafters upon which roofs are installed are decided by roofers.

The above is a general description of each trade that you will come into contact with. If you are new to self-build then it is almost certain that your knowledge of each trade will be rudimentary and it is useful to know the basics at least, in particular if you are hands on and are managing the project yourself, or at least getting involved in elements of the management.

Now read an overview of the Main Points from Chapter 8 overleaf.

****

An Overview of the Main Points from Chapter 8-Main Trades people Involved in Construction

- You will need to establish a good relationship with a builder. Although there are good reputable builders out there, there are also cowboys who can ruin your life! If you intend to manage the project yourself, you also need to establish relationships with sub-contractors.

- In most cases however, you will choose a builder who will act as a main contractor, and will manage the whole process either on a labour only basis, leaving you to purchase the materials, or who will do the lot for you.

- In any event, it is important to understand the roles of the various trades involved in the construction of a house.

Chapter 9

Estimating Building Works

In chapter 2, we discussed the overall costs of self-build, including land and fees and all other associated costs which showed you how to arrive at a total figure in order to be in a position to obtain finance to build your home. In this chapter we will look at the specific build costs of a home, in this case a traditional brick-built 3 bedroom Dorma bungalow. Throughout the book, we have concentrated on construction using traditional methods. There are, of course, timber frame houses and derivations on those, such as prefab construction. However, whichever method you use, the principles for arriving at costs are the same.

Paul and Jayne Marshall.

Using the overall costs of our current project, a 3 bed bungalow, as a typical example of self-build costs outside of London and the South East, in this case in Derbyshire, we have taken the overall construction cost (minus land cost) of £135,000 and broken this down into the labour cost and the contract electrics and plumbing cost totalling £50,000. We then itemised the costs of individual areas which (broadly) follows the sequence of construction

	COST	%
TOTAL BUILD ESTIMATE	**£135,000**	
LABOUR ON THE PROJECT	£40,000	30
ELECTRICS AND PLUMBING	£10,000	7.5

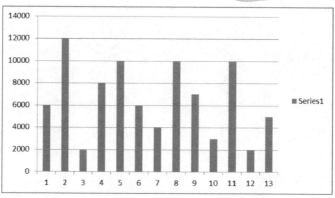

1. Groundworks and drainage	£6,000	4
2. Brick and blockwork	£12,000	8.5
3. First Fix Carpentry	£2,000	1.5
4. Joinery	£8,000	6
5. Roof	£10,000	7.5
6. Plastering	£6,000	4
7. Second Fix carpentry	£4,000	3
8. Externals	£10,000	7.5
9. Kitchen	£7,000	5
10. Fittings and decorating	£3000	3.2
11. Plumbing bathrooms heating	£10,000	7.5
12. Final electrics	£2,000	1.5
13. professionals	£5,000	3.3
OVERALL TOTAL	**£135,000**	**100**

Looking at the figures above, it can be seen that the labour element, at £40,000, is by far the biggest cost of the project. The electrics and plumbing, as a package, at £10,000 brought the initial cost up to £50,000. Following on from this the thirteen elements represented on the chart above brought the overall cost to £135,000. By looking at the costs and percentages, the money allocated to each section of the project and the proportion of overall cost can be clearly identified.

In Chapter ten, we will look at the actual construction process, illustrated by photos of each main area.

****

Chapter 10

The Main Stages of Construction

Stage 1. Choosing and buying your site

In chapter 1, we discussed finding a site. Houses should be built on firm and stable ground so that they don't move, tilt, crack or fall down. Shifting sand and waterlogged earth are bad soils to build on because the ground can move and damage the house. You will need power, water and communication so you need to site your house close to electricity, and telephone cables, water and sewerage mains. You might also consider how close the building is to roads, towns, transport, hospitals, etc.

Below: Site of Paul and Jayne Marshall's proposed self-build at the side of the cottage. Solid land close to water, power, sewerage, telephone etc.

Stage 2. Buying and planning materials and labour

To build your home, you will need various materials, for example concrete for the base, bricks for the walls, roof tiles and so on. You will need to decide what materials to use, where to buy them from and when to get the materials to the building site. It is best to get the materials delivered just before they are needed so the site is not crowded with materials that are not being used for a while.

Stage 3. Levelling the land

Below. Land levelled and footprint marked out, necessary building materials already on site.

The outline of the building (its footprint) is marked according to the architect's plan. The ground in and around the footprint is then cleared of trees, plants, rocks and larger stones. Then earth is moved and flattened to create a horizontal base to build on.

Stage 4. Bringing in utility pipes

On any new site, pipes will need to be put in the ground and connected to supplies to bring in mains water, take away waste water (sewage) bring in electricity, gas, and telephone lines.

Stage 5. Digging/Laying the foundations to form the base of the house

The foundations are the part of a house that is at the bottom of the structure and underground. They are like the roots of a tree – they stop the house from moving sideways or sinking into the ground and they keep the house stable in wind. There are different ways to lay foundations. The method used depends on the style of house, the type of soil the house is being built on and what temperatures the soil is likely to experience. If the soil is likely to freeze, the foundations must be dug to at least 30cm below where the soil freezes (soil gets warmer the deeper down you dig).

Foundations are generally laid on a footing (base), which is wider than the foundation wall. The footing makes sure the weight of the building on top is spread out over the ground. Generally, the bigger, taller and heavier the building, the deeper the foundations need to be to keep the building in place and stop it sinking, tilting or moving. Foundations in light, soft soil need to be wider to prevent them sinking into the soil. The heavier and thicker the soil, the thinner the foundations can be.

Foundations often incorporate a moisture barrier to stop dampness in the ground creeping up the walls and making the house damp. The materials used in the foundations should not rot in the ground. In the past brick and stone were popular but modern buildings generally use concrete as it is quick and easy to pour into trenches in the ground. The concrete can take a few weeks to harden throughout.

Stage 6. Erecting the frame of the building

Some buildings use a wooden frame that goes up before the walls and acts as a skeleton of the house structure for the rest of the build. These are called timber-framed buildings. Before putting up the frame, the position of the walls is set out, leaving gaps for doors and windows.

Larger buildings might use steel struts for a frame instead of wood as steel can carry more weight than wood. It is very important to make sure that the walls are vertical and the window frames and doorframes are the right size.

In some houses (see below) the walls go up without a frame going up first e.g. houses made of stone, brick or concrete. In these houses, the walls go up one layer at a time, with carpenters coming to put joists floor at each floor level before the wall is built further up for the next ceiling/floor level.

Stage 6-Internal construction

As the building is constructed, and the walls are erected, windows and staircases plus floor joists are put in place.

Stage 7. Putting the roof on

Depending on the design of a building, a roof can be put on before or after the walls. The roof trusses sit on a roof plate which is a timber frame on top of the finished brickwork.

See overleaf.

A pitched (sloping) roof needs a framework structure of trusses to support it. A sloping roof is generally used in countries where there is a lot of rain. Usually a thin covering of wood is put over the frame and then roofing felt (tar paper) is put over this to stop water getting in. Finally shingles (roof tiles) are usually laid on top of this. Shingles can be made from terracotta, asphalt, wood, slate or other materials such as copper. The heavier the material, the stronger the supporting frame needs to be.

A flat roof can be made from something as simple as a corrugated sheet of metal or plastic to a more complex layered structure of felt and fibreglass containing asphalt. Flat roofs should slope slightly so water can drain away.

Flat roofs are popular in hot countries where any puddles of rain quickly evaporate.

Stage 8. Initial wiring and plumbing

Before the walls are finished, the wires to electrical sockets, light fittings, ovens, boilers are installed. At this stage the wires are left exposed from the wall.

Water and waste pipes are run to sinks, baths, washing machines. If you want to install telephone lines and gas pipes, this is done at the same time.

Stage 9. Insulating the building

Insulation is usually put between the inside and outside walls, in the roof/top of the house and sometimes in the ground floor.

To save energy (and money) install the maximum recommended amount of insulation.

Stage 10. Putting ceilings in

Many materials can be used to create ceilings but most houses in the UK have plaster ceilings.

Stage 11. Installing plumbing

Showers, sinks, toilets and other large plumbing fixtures should be installed before the walls are finished so the walls are not damaged by putting them in. Water pipes should be connected and tested to make sure water runs to the correct places. Bathtubs should be filled so they settle into their frame with the weight of the water. This should be done before the bathroom is tiled and the corners sealed.

Stage 12. Finishing the inside walls/erecting doors and installing windows

Once the pipes, wires, insulation and plumbing are in place, the inside walls, doors and windows can be finished.

See overleaf.

Stage 13. Install fixtures and decorate

Interior finishing involves:

- connecting exposed electric wires to light fittings, electrical sockets
- installing fixed cupboards, kitchen cabinets and appliances
- decorating with wallpaper, skirting boards, cornices
- painting.

Stage 14. Lay flooring

The final thing to do in the house once all the decorating and painting is done is to lay flooring and/or carpets.

Stage 15 Landscaping

Landscaping around the outside of a building is when any driveways, paths, patios or decks are put in and gardens are completed with trees, shrubs, flowers and grass.

Stage 16. Snagging

Snagging is the final stage when any remaining work is done, problems are addressed and any damage repaired before people start to use the building.

The Defects period

The defects liability period (now called the 'rectification period' in Joint Contracts Tribunal (JCT) contracts) begins upon certification of practical completion and typically lasts six to twelve months.

During this period, the client reports any defects that arise to the contract administrator who decides whether they are defects (i.e. works that are not in accordance with the contract), or whether they are in fact maintenance issues. If the contract administrator considers they are defects, then they may issue instructions to the contractor to make them good within a reasonable time.

NB: It is actually the contractor's responsibility to identify and rectify defects, not the clients, so if the client does bring defects to the contractor's notice, they should make clear that this is not a comprehensive list of all defects

At the end of the defects liability period, the contract administrator prepares a schedule of defects, listing those defects that have not yet been rectified, and agrees with the contractor the date by which they will be rectified. The contractor must in any event rectify them within a reasonable time.

When the contract administrator considers all the items on the schedule of defects have been rectified, they issue a certificate of making good defects. This has the effect of releasing the remainder of any retention and results in the final certificate being issued.

It is important to note that the defects liability period is not a chance to correct problems apparent at practical completion, it is a period during which the contractor may be recalled to rectify defects which appear. If there are defects apparent before practical completion, then these should be rectified before a certificate of practical completion is issued. This can however put the contract administrator in a difficult position, as both the contractor and the client may be keen to issue the certificate (so that the building can be handed over) and yet defects (more than a de minimis) are apparent in the works. Issuing the certificate could render the contract administrator liable for problems that this causes for example in the calculation of liquidated damages.

In practice it is not unusual, particularly if it is in the client's interests, for a certificate of practical completion to be issued with an attached list of minor omissions and defects to be rectified in the defects period. An example of this would be if the certificate of practical completion might trigger tenants fit out and subsequent payment of rent, when it is in nobody's interest to delay the programme just for delivery of a piece of door furniture or a replacement light fitting.

If the contract administrator is pressured to certify practical completion even though the works are not complete, they might consider informing the client in writing of the potential problems of doing so, obtaining written consent from the client to certify practical completion and obtaining agreement from the contractor that they will complete the works and rectify any defects. If the contract administrator is not confident about the potential problems this may cause, they may advise the client to seek legal advise.

NB. Some forms of contract allow for an alternative position of 'substantial completion'.

NNB. On construction management contracts, a separate certificate of practical completion must be issued for each trade contract. This means there may be a number of defects liability periods. The same is true on management contracts, where each works contract must be certified individually.

Guarantees: new-build homes come with a 10-year NHBC warranty covering structural defects. Most developers also provide their own two-year warranty.

****

PART 4

POST CONSTRUCTION AND MOVING INTO YOUR HOME

- The Steps Following Completion of Your Home

Chapter 11

The Steps Following Completion Of Your Home

Completion of the build

This is the stage known as 'practical completion'. The main point is that you as the self-builder need to be satisfied with the completed work. This is the end of a long road for you and it will be your future home, so all needs to be in order.

Completion certificates

Building Control and Building Standards can issue a completion certificate providing the various stages of inspection have gained approval. This certificate will usually be required by your lender, solicitor or if you sell the house on in future. It shows that the main stages of the building works have been completed so far as can be reasonably ascertained by a building inspector. The Benchmark certification system is also included in this.

The certificate does not guarantee that there are no hidden problems and it is not insurance cover. If you were to sell the house after completion then usually this certificate is what the buyer would ask to see, probably along with an up to date surveyor's report.

In Scotland, prior to the occupation of a new property a 'Habitation certificate' needs to be issued by Building Standards. This would be in the form of either a Temporary Occupation/Use Certificate or Certificate of Completion. (It is an offence to occupy a building without one or the other).

Your architect can issue a 'practical completion certificate' when the job is finished. This may be all your lender needs to release a final stage payment of the mortgage. This certificate basically states that the building has been completed according to the plans and is usually based on visits made by the architect during the building work. However, and it is important to note this, It does not imply that there are no hidden defects which might come to light later and it is not the same as insurance cover such as is provided by an NHBC certificate.

Release of final funds

Your lender, if you have one, needs to be satisfied about releasing final payments. They will require both building control and architects completion certificates before they pay the final stage. Sometimes they will rely on a surveyor's report. At this point you will pay off your builder. However, you may want to retain some money, which is the usual practice.

See overleaf for more about retention.

Insurance

The NHBC insurance comes into effect on completion and lasts for ten years. Building site insurance can cease on completion and normal building insurance take over.

VAT reclaim

VAT can probably be claimed on completion. An EPC is required. See chapter 2 for more on VAT and self-build.

Retention money

It is normal, on small domestic scale building contracts, to withhold 5% of the payment due to the builder for a period of 6-months, or occasionally one year, after practical completion. This is usually known as the defects liability period. At the end of the period a schedule of defects is drawn up, the builder makes good any defects and then your architect (if you use one) issues a Certificate of Making Good Defects. The builder is then paid off in full. This all needs agreeing beforehand with the builder, usually in the form of a written contract.

Log book and instructions

Modern houses can be very technical and most self builders will compile a log book which is a guide for users. There are various ways of doing this, one being the maintenance of computer records. Probably the best way is to have all the information collected in a ring binder (along with logs of service intervals etc.) and then to also have individual instruction sheets in laminated covers, discreetly situated near the relevant appliances such as boilers, consumer units, broadband connections, fire alarms etc.

Disputes

Obviously, the better the terms you are on with the various members of the building team, the easier it is to resolve disputes, but if and when they do occur there are various approaches to sorting things out.

Disputes tend to be of two distinct types
- when the builder has done a bad job
- when the design was flawed in the first place. This is usually down to the architect, engineer etc.

If a serious design problem shows up then you could try to sue your architect, engineer etc. They will most likely have professional indemnity to cover this eventuality.

If it is the builder's fault then the retention money is an incentive for him to come back and remedy the work. However this is only intended to address relatively minor problems at the end of the job. If it is after the retention period then you must rely on the good will of the builder, and, failing that, take the builder to court.

However it is not usually as clear cut as all this, especially around disputes with builders. The first thing to do is consult the documents which make up the contract. These will probably involve drawings (some of which have been approved by Building Control and have a legal basis), specifications which relate to the drawings and any other documents and notes which may have been agreed as work has proceeded. Verbal agreements can be difficult to pin down.

Mediation and conciliation

Rather than taking a builder to court or to arbitration it may be sufficient to bring in a mediator. Their job is to take the heat out of the situation and suggest ways of resolving a problem. Not only technically but also psychologically it can make a big difference.

They may be a building professional such as an architect, clerk of works, self build consultant etc. They don't have to be an expert in the field: they may simply bring in outside experts. The support of a building inspector may also be useful but remember that their ambit is restricted to the parts of the work covered by the Building Regulations.

Arbitration

Another alternative to litigation is arbitration and it is often included as a part of the building contract. The arbitrator will be an expert in construction, which may be a better situation than with a general judge in a court. Both parties have to agree on arbitration and also on who the arbitrator (or appointing body) is. The arbitrator's decision is then binding. It can only be challenged in court over a point of law.

Litigation

Taking a builder to court is a civil law matter. The NBS https://www.thenbs.com have a useful section on legislation along with arbitration, mediation and conciliation.

****

An overview of the Main Points from Chapter 11-Steps Following Completion of your Home

- This is the stage known as 'practical completion'. You as the self-builder need to be satisfied with the completed work. This is the end of a long road for you and it will be your future home, so all needs to be in order.

- Building Control and Building Standards can issue a completion certificate providing the various stages of inspection have gained approval. This certificate will usually be required by your lender, solicitor or if you sell the house on in future

- Your architect can issue a 'practical completion certificate' when the job is finished. This may be all your lender needs to release a final stage payment of the mortgage.

- The NHBC insurance comes into effect on completion and lasts for ten years. Building site insurance can cease on completion and normal building insurance take over.

- Any resultant disputes should preferably be resolved either directly with the builder or through Arbitration or Mediation.

Conclusion to This Book

Hopefully, after reading this book you will have gained a clearer idea of the processes involved in self-building a home. We have, of necessity, kept it as brief as possible. To include every stage in full would need a book which would be both unwieldy and confusing. Our aim has been to discuss and summarise each area, which provides a series of steps and a solid background for those who want to go forward and build a home. The actual practice of carrying out a self-build, from A to Z will complete the learning curve.

The A to Z comprises the following:

1. Understanding self-build, the reasons why you would want to self-build and the role of government in facilitating self-build.

2. Finding a site-how to go about finding a suitable plot.

3. Working out the costs of building your own home.

4. Raising finance to build your own home.

5. Understanding what kind of home you would like and also the limitations. Working with an Architect.

6. Obtaining planning permission to build your home.

7. Understanding building control and the role of the local authority.

8. Understanding the construction process and estimating costs of construction and stages of construction.

9. Understanding of the steps after construction and prior to moving in.

We (Paul and Jayne) have carried out a number of successful self-builds and have found the whole process an adventure. Taking on such a task is not for the faint hearted but it is within the grasp of most people who take an interest in this route to finding a home, at a cost within their reach.

Good luck in your venture!

****

Useful websites

Most of the websites listed below contain a wealth of general information concerning self-build, in addition to the specific areas listed. We the authors do not endorse any particular website or company.

Finding a Plot of Land

www.selfbuildportal.org.uk/finding-a-plot
(www.righttobuildportal.org)
This website is produced by the National Custom & Self Build Association (NaCSBA) and is endorsed by the Government.

www.self-build.co.uk/finding-plot
This website is allied to Build It and is a very useful place to start when looking for a plot of land.

www.plotfinder.net
The UK's leading land and renovation finding service - from Homebuilding & Renovating. A subscription based website.

www.buildstore.co.uk/findingland
Advertises itself as the UK's largest and most accurate database of genuine Self-Build Land, Renovation & Conversion opportunities

www.plotbrowser.com

This is a free online plot finding service

Costs of Self-Build

www.homebuilding.co.uk

The site of homebuilding and Renovating, this has very useful information relating to the costs of self-build and contains a build-cost calculator.

www.self-build.co.uk/costs/

Contains a wealth of information relating to the costs of building a home-aimed at self-builders

www.jewson.co.uk

A building supplies company-Jewsons has its own cost calculator, which is also used by other sites.

www. selfbuildportal.org.uk

Contains useful information relating to budgeting and costs.

Obtaining a Self-Build Mortgage

www.buildstore.co.uk/finance/selfbuildmortgages.

Contains a wealth of information concerning raising mortgage finance for self-build projects.

www.homebuilding.co.uk › Finance

Similarly contains comprehensive financing information

www.moneysupermarket.com/mortgages/self-build

Contains useful information about various mortgage providers.

www.money.co.uk/mortgages/self-build-mortgages

A useful comparison website which offers information about various self-build mortgage providers and their rates.

mortgageadvisers.which.co.uk

The website of Which? the consumer group this offers a self-build mortgage service.

The Design Process

www.mapletimberframe.uk/self-build/timber-frame

A Site for self-builders specialising in timber frame houses

Finecomb.com/Build A House Online

Online design and build site

www.selfbuildanddesign.com

Practical advice and inspiration for all stages of a self-build project whether it is a new build, redevelopment, extension, conversion, modernisation or home.

www.self-build.co.uk
Online site-home of Build-it Magazine-offers
comprehensive design advice to all self builders.

www.potton.co.uk
Timber frame specialist offers design and build advice

Obtaining Planning Permission

www.gov.uk/browse/housing-local-services/planning-permission
Government site offering comprehensive advice relating to
all aspects of planning permission

www.planningportal.co.uk
Online site offering advice and guidance concerning
planning permission

*www.homebuilding.co.*uk › *Planning permission*
Explains the planning process and associated costs in
detail.

Building Regulations and Control

www.self-build.co.uk/guide-building-regulations
Offers comprehensive advice and guidance concerning
building regulations and building control

https://www.planningportal.co.uk
As above, offers comprehensive advice on building
regulations for self-builders

The Construction Process

The following sites all offer comprehensive advice concerning the construction process:

www.self-build.co.uk
www.homebuilding.co.uk
www.selfbuildportal.org.uk

(The Self-Build Portal is a Government-endorsed website for the would-be self builder - providing impartial advice and useful information on building your home).

www.acarchitects.biz/self-build-blog-guide-to-construction-methods
UK Self Build Architects, ACA, offer a guide to the most popular construction methods used in the Self-Build sector today.

Self-build contracts
www.the-self-build-guide.co.uk/construction-contracts
Offer comprehensive advice concerning construction contracts for self builders

https://www.contractstore.com/construction/self-builders
Offers very useful contract templates and advice for self builders.

https://www.contractstore.com/construction/self-builders
Offers building contract templates for developer or home owner appointing contractor to build a new house extension or other construction work.

Completion Of Your Home

www.selfbuild-central.co.uk/construction/on-completion
Offers comprehensive advice on the process of practical completion of a home

thegreenhome.co.uk/.../a-guide-to-self-build-completion-certificates-
Offers some advice on the self-build completion certificates you will need to acquire at the end of your building project.

**

Index

Adjudication, 130
Architects, 11, 76, 127
Architects' Registration Board, 76
Areas of Natural Beauty, 96
Arrears-type self build mortgage, 68
Auctioneers, 37

Benchmark certification system, 163
Bespoke designs, 6, 77
Bespoke home, 13
Bricklayers, 134
Brownfield sites, 4, 24, 25
Builders merchants, 57
Building Contract, 7, 122
Building Control, 7, 75, 113, 114, 163, 167, 169, 175
Building control department, 113, 117
Building materials, 5, 56, 61
Building Regulations, 77, 92, 113, 114, 117, 167

Carpenters, 8, 135, 136
Certificate of Completion, 114, 117, 164
Changes of use, 116
Completion, 8, 9, 126, 163, 177
Construction costs further, 56
Construction management, 7, 121
Conversions, 5, 61
Conveyancing, 46
Costs of Self-Build, 5, 51, 173
Custom build schemes, 32

Defective Premises Act 1972, 123, 127
Defects Liability, 8, 126
Delays, 8, 125

Demolition, 54
Design and Access Statement, 97
Designing Your Home, 6, 75
Developers, 36
Disabled access, 98
Dispute resolution, 123, 131
Disused land, 38
Do-It-Yourself design, 6, 77

Electricians, 8, 137
Energy performance, 116
Energy saving, 4, 13
Environment Agency, 100
Estate agents, 32
Estimating Building Works, 8, 141
Extending planning permission, 7, 101

First fix, 67
Flipping, 36
Floor plan, 80
Freehold, 45
Full planning permission, 6, 96, 100

Garden grabbing, 4, 25
General Conditions of Sale, 37
Green belt, 25
Greenfield sites, 4, 25
Ground, 45, 80

Groundworker, 133

Habitation certificate, 164
Help to Build, 16, 17, 18
Housing and Planning Act 2016, 15, 39

Individual building plots, 23
Insurance, 8, 9, 122, 127, 131, 165

Land Registry, 48, 49
Leasehold, 45
Liquidated Damages, 8, 127
Litigation, 9, 168
Local connection, 5, 41
Local Planning Authority, 101

Mediation, 9, 130, 167

National Custom and Self-Build Association (NaCSBA), 14
NHBC certificate, 164
Northern Ireland, 16, 41, 93, 95, 96

Outline planning permission, 96, 111

Painters, 8, 138
Permitted Development Rights, 6, 95
Planning application requirements, 7, 97
Planning Conditions, 7, 97, 98
Planning decisions, 7, 98
Plasterers, 8, 136
Plot hunting, 38
Plumbing, 137
Professional fees, 56
Project manager, 56
Property auctions, 34
Property investors, 35
Property scans, 33
Purchase of a plot, 52

Qualified Architectural Technologists (MCIAT, 76
Quantity surveyor, 56

Reclaiming VAT, 5, 60
Registered Social Landlord, 15
Repossessions, 35
Residential plumbers, 137
Retention money, 9, 165
Right of access, 45
Right to Build, 4, 5, 15, 16, 24, 39, 40, 41, 42
Roofers, 8, 138
Royal Institute of British Architects (R.I.B.A.), 76
Royal Institution of Architects in Scotland, 76

Self-Build and Custom Housebuilding Act 2015, 15, 39
Self-build finances, 51
self-build mortgage, 43, 65, 174
Self-build mortgage, 55
Serviced plots, 15, 32, 39, 50
Special Conditions of Sale, 37
Stage payments, 67
Stamp Duty Land Tax, 5, 53
Structural engineer, 56
Subcontractors, 133

The Chartered Institute of Building, 56
The Damp proof course, 134
The Land Bank Partnership, 30, 50
The Planning Portal, 101
Time Extensions, 8, 125

Value added tax (VAT), 5, 59
Vehicle access, 38

****

Appendix 1 Overleaf.

Notice of Planning Consent 3 Bed Dorma Bungalow (Paul and Jayne Marshall).

www.erewash.gov.uk **EREWASH**

Resources Directorate, Planning & Regeneration
Town Hall Ilkeston
Derbyshire DE7 5RP
Switchboard: 0115 907 2244

Mr Paul Gaughan
Unit E14, Langham Park
Lows Lane
Stanton By Dale
Derbyshire
DE7 4RJ

TOWN AND COUNTRY PLANNING ACT 1990

NOTICE OF DECISION

Part 1: Applicant Details

Applicant: **MR MARSHALL**

Application Ref: **ERE/0615/0023**

Proposal: **ERECTION OF A DETACHED THREE BEDROOM CHALET
                    BUNGALOW WITH ROOMS IN THE ROOF SPACE**

Site Address: **LAND TO REAR OF, 68 & 70 STATION ROAD, STANLEY,
                    DERBYSHIRE, DE7 6FB**

Part 2: Decision

Erewash Borough Council in pursuance of powers under the above mentioned Act
hereby

GRANT PERMISSION

for the development in accordance with the application, subject to compliance with
the condition(s) imposed (in Part 3 below), and the subsequent approval of all
matters referred to in the conditions:

Part 3: Condition(s)

1. The development shall be begun before the expiration of three years from the date of this permission.

Reason
To comply with the requirements of Section 91 of the Town and Country Planning Act 1990.

2. This permission relates to drawings: Site Location Plan (1:1250) and Drw Nos. DB/PJM/14/049/03 Revision A (Proposed Site Block Plan) & DB/PJM/14/049/02 (Proposed Floor Layouts and Elevations), all validated on the 11th June 2015. Any variation to the approved drawings may need the approval of the local planning authority.

Reason
For the avoidance of doubt as to what is approved.

3. Prior to any other works commencing, the entire site frontage shall be cleared, and maintained thereafter clear, of any obstruction exceeding 1m in height (600mm for vegetation) relative to the road level for a distance of 2.4m from the nearside carriageway edge in order to maximise the visibility available to drivers emerging onto the highway.

Reason
In the interests of highway safety.

4. The proposed dwelling shall not be occupied until space has been provided within the application site in accordance with the revised application drawings for the parking of residents' vehicles, laid out, surfaced and maintained throughout the life of the development free from any impediment to its designated use.

Reason
In the interests of highway safety.

5. Notwithstanding the provisions of Part 2, Schedule 2 of the Town and Country Planning (General Permitted Development) (England) Order 2015 (or any Order revoking and re-enacting that Order), no gates or other barriers shall be erected across the entire frontage of the application site unless planning permission has first been granted by the local planning authority.

Reason
To allow for easy access and egress in the interests of highway safety.

6. The construction of the dwelling shall not commence until samples of the proposed materials to be used in the external construction of the development, including areas of hardsurfacing, have been submitted to, and approved in writing by the Local Planning Authority and the development shall only be undertaken in accordance with the materials so approved and shall be retained as such thereafter.

Reason
To ensure a satisfactory standard of external appearance.

7. The proposed dwelling shall not be occupied until a detailed scheme for the boundary treatment of the site, including position, design and materials, and to include all boundaries or divisions within the site, has been submitted to and approved in writing by the local planning authority. The approved scheme shall be completed before the dwelling is first occupied or such other timetable as may first have been agreed in writing with the local planning authority.

Reason
To preserve the amenities of the occupants of nearby properties and in the interests of the visual amenity of the area.

8a) The development shall not commence until a scheme to identify and control any environmental risk is developed and undertaken. This will include a desk top study (Preliminary Risk Assessment / Phase I Investigation) and, if indicated by the desk top study, an intrusive investigation (Generic Risk Assessment/ Phase II Investigation). The desk top study and then the scope of the intrusive investigation must be approved in writing by the local planning authority before commencement. In reaching its decision to approve such proposals the local planning authority will have regard to currently pertaining government guidance as set out in the CLR series of documents (particularly CLR 11) issued by DEFRA or any subsequent guidance which replaces it.
b) A written method statement detailing the remediation requirements to deal with any environmental risks associated with this site shall be submitted to and approved in writing by the local planning authority prior to commencement of the remedial works. The method statement should also include details of all works to be undertaken, proposed remediation objectives and remediation criteria, timetable of works and site management procedures. All requirements shall be implemented according to the schedule of works indicated on the Method Statement and completed to the satisfaction of the local planning authority prior to the development being brought into use. No deviation shall be made from this scheme without the express written agreement of the local planning authority.
c) If, during the development, any contamination is identified that has not been considered previously, then, other than to make the area safe or prevent environmental harm, no further work shall be carried out in the contaminated area until additional remediation proposals for this material have been submitted to and approved in writing by the local planning authority. These proposals would normally involve an investigation and an appropriate level of risk assessment. Any approved proposals shall thereafter form part of the Remediation Method Statement.
d) Prior to the development first being brought into use a verification report must be submitted to the Local planning authority demonstrating that the works have been carried out. The report shall provide verification that the remediation works have been carried out in accordance with the approved Method Statement. The development should not be brought into use until the verification report has been submitted to and approved in writing by the local planning authority.
e) In the event that it is proposed to import soil onto site in connection with the development the proposed soil shall be sampled at source such that a representative sample is obtained and analysed in a laboratory that is accredited under the MCERTS Chemical testing of Soil Scheme or another approved scheme the results of which shall be submitted to the local planning authority for consideration. Only the soil approved in writing by the local planning authority shall be used on site.

Reason

To ensure that risks from land contamination to the future users of the land and neighbouring land are minimised, together with those to controlled waters, property and ecological systems, and to ensure that the development can be carried out safely without unacceptable risks to workers, neighbours and other offsite receptors.

9. No development shall take place until the scheme of intrusive site investigations as detailed in the submitted Coal Mining Risk Assessment dated May 2015 has been undertaken. A report detailing the findings arising from the site investigation shall be submitted to the Local Planning Authority before any development commences. If any land instability issues are found resulting from, for example, past mining activity during the site investigation, a report specifying the measures to be taken to remediate the site to render it suitable for the development hereby permitted shall be submitted to and approved in writing by the Local Planning Authority. The site shall be remediated in accordance with the approved measures before development begins.

Reason

To ensure that risks from land instability and mining related hazards to the future users of the land and neighbouring land are minimised, and to ensure that the development can be carried out safely without unacceptable risks to workers, neighbours and the general public.

10. The hours of working on the construction of the development, and deliveries to/collection from the development site shall only take place between the hours of 7.30am and 6.00pm on Monday to Friday; 8.00am and 1.00pm on Saturday with no working taking place on Sundays, Bank and Public Holidays.

Reason
In the interests of residential amenity.

Part 4: Positive and proactive statement

There were no problems for which the Local Planning Authority had to seek a solution in relation to this application.

Part 5: Notes to applicant

1. Pursuant to Sections 149 and 151 of the Highways Act 1980, steps shall be taken to ensure that mud or other extraneous material is not carried out of the site and deposited on the public highway. Should such deposits occur, it is the applicant's responsibility to ensure that all reasonable steps (e.g. street sweeping) are taken to maintain the roads in the vicinity of the site to a satisfactory level of cleanliness.

2. Due to the location of the site within a residential area the applicant should take all reasonably practicable steps to minimise noise and dust nuisance that may arise from activities on site. The applicant is also advised that material arising from the clearance of vegetation should not be burnt on site in order to prevent smoke nuisance being caused to neighbouring properties.

Date: 14 August 2015   Signed _____
        Steve Birkinshaw
        Head of Planning & Regeneration

ATTENTION IS CALLED TO THE NOTES BELOW

Discharge of Conditions fees:
http://www.planningportal.gov.uk/uploads/english_application_fees.pdf

Appeals to the Secretary of State

- If you are aggrieved by the decision of your local planning authority to refuse permission for the proposed development or to grant it subject to conditions, then you can appeal to the Secretary of State under section 78 of the Town and Country Planning Act 1990.

- If you want to appeal against your local planning authority's decision then you must do so within 6 months of the date of this notice.

- Appeals must be made using a form which you can get from the Secretary of State at Temple Quay House, 2 The Square, Temple Quay, Bristol BS1 6PN (Tel: 0303 444 5000) or online at www.planningportal.gov.uk/planning/appeals

- The Secretary of State can allow a longer period for giving notice of an appeal but will not normally be prepared to use this power unless there are special circumstances which excuse the delay in giving notice of appeal.

- The Secretary of State need not consider an appeal if it seems to the Secretary of State that the local planning authority could not have granted planning permission for the proposed development or could not have granted it without the conditions they imposed, having regard to the statutory requirements, to the provisions of any development order and to any directions given under a development order.

APPLICATION REFERENCE: ERE/0615/0023

ADDRESS:
Land to Rear of
68 & 70 Station Road
Stanley
Derbyshire

DESCRIPTION

ERECTION OF A DETACHED THREE BEDROOM CHALET BUNGALOW WITH
ROOMS IN THE ROOF SPACE

PROPOSALS

Full planning permission is sought to erect a detached three bedroom chalet bungalow
with rooms in the roof space at the rear of Nos. 68 and 70 Station Road, Stanley. The
main body of the proposed dwelling will measure approximately 14m (w) by 9m (d) with
a height of 6.4m (2.5m high to the eaves). An attached garage will project 5.5m from the
front elevation and a single storey orangery will project 4.05m from the rear elevation.
The rooms in the roof space will be served by two rear dormers and front & rear roof
lights. The main body of the proposed dwelling will be sited approximately 4m from the
rear elevation of No.70 Station Road, 5m from the boundary with No.64 Station Road,
1.1m from the boundary with No.92 Station Road and 13m from the rear boundary.

The proposed dwelling will provide approx. four off-street parking spaces and will be
accessed via a shared driveway to the eastern side of Nos. 68 & 70.

SITE AND SURROUNDINGS

The site is located within the village envelope and currently forms part of the rear garden
to No.70 Station Road. The area contains a mix of house types in a range of
orientations. The site, other than the shared driveway, is sited approximately 21m away
from the back edge of the public highway. Boundary treatment consists of approx 1.6m
– 1.8m high fencing plus planting to the sides and rear boundaries.

The application site falls within a referral area in relation to historic coal mining activity in
the area.

RELEVANT SITE HISTORY

0684/0056 - Erection of detached bungalow - Approved

0205/0033 - Erection of detached bungalow - Refused on highway grounds and
backland development - Appeal Dismissed

0213/0026 - Erection of detached bungalow – Refused on highway grounds
(intensification of use of a substandard access, restricted visibility & potential danger to
other highway users)

POLICY CONTEXT

National Policy

National Planning Policy Framework (NPPF)

Erewash Core Strategy

Policy A: Presumption in Favour of Sustainable Development
Policy 2: The Spatial Strategy
Policy 8: Housing Size, Mix and Choice
Policy 10: Design and Enhancing Local Identity

Erewash Saved Policies

H3: Village Housing Development
H12: Quality and Design
DC1: Backland and Tandem development

Supplementary Planning Documents

Erewash Borough Supplementary Planning Document (SPD) 2006: Design

CONSULTATIONS

Ward Councillors – No representations received.

Stanley and Stanley Common Parish Council – No objection in principle to the proposed bungalow and backland siting subject to neighbour consultation. The proposed access arrangement for Nos.68/70 however appears to remove the current open fronted access arrangement and forwards entrance and exit vehicle access for these existing properties and the Parish Council would consider the application proposals an undesirable and retrograde step given the nearby sharp bend in the highway and actual egress visibility conditions existing here.

DCC Highways – Comments as follows:
- The Highways Authority has been involved in pre-application discussions with the agent. Consistently raised concerns over achievable visibility and intensification of use of existing access but unlikely to recommend refusal for an additional dwelling should the former access to the western side be re-opened under permitted development subject to the sites frontage being cleared to maximise visibility and space for parking and manoeuvring being provided. Whilst this would not be ideal, the eastern access would continue to serve two dwellings.
- Concerns over the applicant's/agents methodology/reasoning used in arriving at the achievable visibility from the existing access.
- Mitigation to offset traffic impact of proposed development.
- Both accesses fall below modern criteria in terms of emerging visibility.
- The eastern access will be shared by No.70 and the proposed dwelling so will provide access for two dwellings.
- Applicant is able to offer improvements by securing a parallel sightline 2.4m back from the nearside carriageway edge across the entire site frontage together with

2m x 2m x 45 degree pedestrian intervisibility splays on either side (where achievable).

- No highways objections subject to conditions requiring a unilateral undertaking to be entered into to ensure that the western vehicular access and parking area be utilised by residents of the existing property and maintained throughout the life of the development, space to be provided for construction workers, vehicles, storage of plant and materials etc, clearance of the site frontage to maximise visibility, and removal of permitted development rights for gates or other barriers across the site frontage.

In addition to the above, the Council sought further clarification on the above comments to which further comments were subsequently received to clarify that the Highways Authority is not saying that the residents must use the western access/parking, but that it should be provided and maintained available.

Severn Trent Water – No representations received.

Coal Authority – Part of the application site falls within the defined Development High Risk Area. Concur with the recommendations of the Coal Mining Risk Assessment Report. No objection subject to a condition requiring intrusive site investigations to be carried out prior to development.

EBC Environmental Health – No objection subject to conditions concerning hours of construction, dust, site clearance and contaminated land.

REPRESENTATIONS

Adjoining neighbours were notified of the proposals and a site notice was erected. Representations received from the occupiers of Nos. 64 & 92 Station Road. Comments as follows:

No.64 Station Road:
- The site location map shows a brook that runs along the bottom of the site but this is not specified on the application form.
- Proposal previously refused twice on the grounds of unsuitable site access.
- The road is not a slow speed or lightly trafficked rural lane as outlined in the submitted transport report.
- There have recently been a few accidents nearby on the road.
- The proposal seems to be trying to shoehorn a large 3 bed, garage and two storey bungalow into a back garden of an already extended property which is unsuitable for such a development.
- The two storey nature of the proposed property may affect the privacy of any neighbouring properties on all sides.

No.92 Station Road:
- No objection to the principle of the development but have a number of objections to the current proposal.
- The plan does not indicate the existence of our sewer which runs across the land in question. We fear this development poses a serious risk to our sewer.
- The proposed property is extremely close to our garden boundary and the very high roof line would block light to our fruit & veg garden and would feel oppressive.

- Feel that the proposed property could be more centrally located on what is a large plot of land.
- Dispute the evidence in the transportation plan as the traffic flow and parking surveys were carried out well outside peak time hours. In reality the traffic flow and level of on-street parking is far higher than assumed in the report.
- The plans show a turning area for the new property but none for No.70.
- We would not be against a new building on the plot if it had a lower roof line, is further away from our boundary, and does not compromise our sewer.

ASSESSMENT

The main issues for consideration in the determination of this application are considered to be:

- The principle of the development;
- The design of the proposal and impact on the character of the area;
- The impact of the proposed dwellings on the amenity of residents;
- Highway safety;
- Other matters.

The principle of the development

The site is located within the village envelope and is close to local services and public transport. The site is currently surplus garden land and subject to the details of the application it is the type of site that the NPPF supports as being appropriate for development. In addition the Saved Policy H3 supports this type of small-scale housing development within the village of Stanley subject to matters of design, access and location. Accordingly the principle of the development is considered to be acceptable and appropriate in this instance and accords with the NPPF, Core Strategy Policies A and 2, and Policy H3 of the Saved Policies, subject to consideration of the following detailed matters.

The design of the proposal and impact on the character of the area

The siting of the proposed dwelling to the rear of Nos. 68 & 70 ensures that whilst it will be visible from Station Road, it will not have a street presence. Its divergence from the character of the area which is of period two and three storey dwellings, with a front to back orientation generally with a road frontage, is contrary to the general character of the area. However there are examples of historical backland development within the area, and as such the design and character of the proposed dwelling is not considered to be so alien as to warrant a refusal of the planning application. It is therefore considered that the proposal will not present significant harm to the character of the area or the street scene as a result of its backland location. Whilst a planning application in 2005 was refused and the appeal dismissed on the grounds of backland development it is considered that since that time the general presumption of planning policy has shifted such that there is a presumption in favour of sustainable development and the Council heavily relies on this type of site coming forward to uplift its housing provision. The development is on balance considered to be acceptable in relation to Saved Policy DC1 as whilst it does not reflect the general pattern of development it does not result in significant harm to the character of the area with limited visibility within the street scene available. The amenity space provided for the dwelling is considered to be of sufficient size and will provide an acceptable level of amenity for future occupiers. The proposal is therefore considered to comply with the policy requirements relating to design and local identity/character within Policy 10 of the Core Strategy and Policy H12 of the Saved Policies.

The impact of the proposal on nearby residents

The proposed dwelling will be within the 45-degree line of vision from the rear windows of Nos. 64 & 92 Station Road and directly to the rear of Nos. 68 & 70 Station Road. However, this would be at a distance of some 9m from No.64, 15m from No.92, and 10m from Nos. 68 & 70 at its closest point (15m to the main body of the proposed dwelling). Whilst there would be some loss of outlook experienced to these properties, the resultant harm to the occupiers' amenity is not considered sufficient enough to warrant refusal of the application on this ground, particularly when noting the distance to the rear windows of these properties. The proposed dwelling will be set in 6m from the boundary with No.64 to the western side. This, together with the separation distance to that property (9m), is considered sufficient to ensure no undue impact on the residential amenity of the occupants of No.64 through loss of outlook, loss of light or overbearing impact. The proposed dwelling will be set in 1.1m from the boundary with No.92 to the eastern side and approx. 15m from the rear elevation of this property. Concerns have been expressed by the occupiers of No.92 that the building would be overbearing and result in a loss of light to the rear garden. When noting the distance to the rear elevation of that property it is considered that the proposal would not be unduly overbearing. The siting of the proposed dwelling adjacent to the boundary will create an element of overshadowing to the rear garden area of No.92, however any loss of light will be contained towards the rear of the garden and well away from the existing dwelling and as a result the resultant harm to the occupiers amenity through overshadowing / loss of light is not considered so severe as to warrant refusal of the application. Similarly it is not considered that the proposed dwelling would present significant harm to No.96 Station Road which is sited on the opposite side of the rear garden of No.92. It is therefore considered that the proposed dwelling will not cause significant detriment to the amenity enjoyed by the occupiers of neighbouring properties through a loss of outlook, loss of light or overbearing impact.

The proposed ground floor side windows to the western side elevation are not considered to present unreasonable privacy concerns through overlooking given that they will be set well away from the side boundary with No.64 whilst no windows are proposed to the eastern side elevation. The ground floor front windows and the front facing roof lights are not considered to present unreasonable privacy concerns to neighbouring properties to both sides or the front. The separation distance between the proposed rear dormer windows within the roof, and the rear elevations of neighbouring properties to the rear, will be approx. 25m. The proposed dormer windows will also be sited approx. 13m-16m from the angled rear boundary of the site. The separation distances proposed are considered sufficient to prevent significant harm through overshadowing or overbearing impact whilst the proposed dormer windows are not considered to result in a significant loss of privacy through a degree of overlooking above that ordinarily expected in a residential environment, particularly when noting the proposed separation distances. Similarly, it is not considered that the proposed rear windows would result in a significant loss of privacy to neighbouring properties to both sides.

Access to the site is obtained from Station Road alongside Nos. 68 & 70 and adjacent to some of the principal windows to both of these dwellings. Of concern is the impact of this drive adjacent to these neighbouring property windows which has the potential to result in undue noise and disturbance to the current and future occupiers of these dwellings. However as the proposal serves only one additional dwelling it is considered that these movements would be limited and whilst there will be some additional traffic movements again it is considered that this will not amount to significant detriment to the occupiers of Nos. 68 & 70 Station Road.

Accordingly, the proposal is considered to comply with Policy 10 of the Core Strategy and Policy H12 of the Saved Policies which seek to ensure that the amenity of neighbouring occupiers is not adversely affected.

Highway safety

The proposed development will be accessed via the existing vehicular to the eastern side of Nos. 68 & 70. DCC Highways Authority have raised concerns over the applicants methodology/reasoning used in arriving at the achieved visibility from the existing access to Station Road, but have noted that the proposal offers improvements to existing visibility and that the re-use of the vehicle access / parking space to the western side of Nos. 68 & 70 would offset the resultant intensification of the eastern access created by the proposed dwelling. They have therefore raised no objection subject to a unilateral undertaking being entered into to ensure that the western vehicular access and parking area be maintained throughout the life of the development, and subject to conditions requiring the access, visibility splays, and parking spaces to be provided in accordance with the application drawings. In response to this a site visit confirmed that the vehicular access / parking space to the western side is in-situ and would appear to be an established access (albeit one that has previously been fenced off but now re-opened) that benefits from an established dropped kerb. Whilst the previous application was refused on highway safety grounds, it is noted that the western access did not form part of the considerations as at that time the access was fenced off from the highway. However, since this time the fence has been replaced with a gate and the access has been re-opened. It is therefore considered reasonable to consider its presence and availability for use by the occupiers of No.68 when assessing the highways impacts of the current application, notwithstanding its location outside of the application site, and in this regard the Highways Authority have not objected to the current proposal for the intensification of the use of the eastern access subject to the western access being provided and maintained available. Given that the western access is in-situ and available for use, it is considered that the application should be assessed on its merits at the time of the application. In this regard it is not considered necessary to enter into a unilateral undertaking to ensure that this is provided. In addition to the above it is considered that sufficient off-street parking will be provided by the proposed development whilst sufficient space will also be available for the parking of vehicles to the side of No.70. The concerns raised by neighbours and the Parish Council in relation to parking provision, access, visibility, turning area and the details contained in the transportation report are duly noted. However, it is noted that national planning policy, in the form of the NPPF, now specifies that developments should only be refused on highway safety grounds where the impacts of the development are severe. As the Highways Authority have not identified severe harm from these proposals, it is concluded that the concerns over highway access do not constitute justified reasons for refusal.

Other matters

The comments of the Council's Environmental Health Officers relating to contaminated land and the control over the hours of construction, noise and dust, can be embodied in conditions or notes to the applicant, as can the comments of the Coal Authority in relation to intrusive site investigations.

The comments received from the occupiers of Nos. 64 & 92 in relation to the brook and potential impact on the sewer are noted. The impact of the development on any sewers that pass beneath the site is not a material planning consideration and such details will be assessed separately as part of the building regulations application. It is considered that the narrow watercourse shown on the site location plan will be unaffected by the proposed development due to its location approximately 9m away from the closest point of the proposed dwelling.

Conclusion

The proposed development of the site for a three bedroom chalet bungalow is within a sustainable location; the plot is of an appropriate size; the scale and design of the dwellings are considered acceptable; the development would not have an adverse impact on the residential amenity of neighbouring properties and the proposal is not considered to result in severe detriment to highway safety. Accordingly the development is considered to be acceptable.

RECOMMENDATION   APPROVE

CONDITIONS & REASONS

1. The development shall be begun before the expiration of three years from the date of this permission.

Reason
To comply with the requirements of Section 91 of the Town and Country Planning Act 1990.

2. This permission relates to drawings: Site Location Plan (1:1250) and Drw Nos. DB/PJM/14/049/03 Revision A (Proposed Site Block Plan) & DB/PJM/14/049/02 (Proposed Floor Layouts and Elevations), all validated on the 11th June 2015. Any variation to the approved drawings may need the approval of the local planning authority.

Reason
For the avoidance of doubt as to what is approved.

3. Prior to any other works commencing, the entire site frontage shall be cleared, and maintained thereafter clear, of any obstruction exceeding 1m in height (600mm for vegetation) relative to the road level for a distance of 2.4m from the nearside carriageway edge in order to maximise the visibility available to drivers emerging onto the highway.

Reason
In the interests of highway safety.

4. The proposed dwelling shall not be occupied until space has been provided within the application site in accordance with the revised application drawings for the parking of residents' vehicles, laid out, surfaced and maintained throughout the life of the development free from any impediment to its designated use.

Reason
In the interests of highway safety.

5. Notwithstanding the provisions of Part 2, Schedule 2 of the Town and Country Planning (General Permitted Development) (England) Order 2015 (or any Order revoking and re-enacting that Order), no gates or other barriers shall be erected across the entire frontage of the application site unless planning permission has first been granted by the local planning authority.

Reason
To allow for easy access and egress in the interests of highway safety.

6. The construction of the dwelling shall not commence until samples of the proposed materials to be used in the external construction of the development, including areas of hardsurfacing, have been submitted to, and approved in writing by the Local Planning Authority and the development shall only be undertaken in accordance with the materials so approved and shall be retained as such thereafter.

Reason
To ensure a satisfactory standard of external appearance.

7. The proposed dwelling shall not be occupied until a detailed scheme for the boundary treatment of the site, including position, design and materials, and to include all boundaries or divisions within the site, has been submitted to and approved in writing by the local planning authority. The approved scheme shall be completed before the dwelling is first occupied or such other timetable as may first have been agreed in writing with the local planning authority.

Reason
To preserve the amenities of the occupants of nearby properties and in the interests of the visual amenity of the area.

8a) The development shall not commence until a scheme to identify and control any environmental risk is developed and undertaken. This will include a desk top study (Preliminary Risk Assessment / Phase I Investigation) and, if indicated by the desk top study, an intrusive investigation (Generic Risk Assessment/ Phase II Investigation). The desk top study and then the scope of the intrusive investigation must be approved in writing by the local planning authority before commencement. In reaching its decision to approve such proposals the local planning authority will have regard to currently pertaining government guidance as set out in the CLR series of documents (particularly CLR 11) issued by DEFRA or any subsequent guidance which replaces it.
b) A written method statement detailing the remediation requirements to deal with any environmental risks associated with this site shall be submitted to and approved in writing by the local planning authority prior to commencement of the remedial works. The method statement should also include details of all works to be undertaken, proposed remediation objectives and remediation criteria, timetable of works and site management procedures. All requirements shall be implemented according to the schedule of works indicated on the Method Statement and completed to the satisfaction of the local planning authority prior to the development being brought into use. No deviation shall be made from this scheme without the express written agreement of the local planning authority.
c) If, during the development, any contamination is identified that has not been considered previously, then, other than to make the area safe or prevent environmental

harm, no further work shall be carried out in the contaminated area until additional remediation proposals for this material have been submitted to and approved in writing by the local planning authority. These proposals would normally involve an investigation and an appropriate level of risk assessment. Any approved proposals shall thereafter form part of the Remediation Method Statement.

d) Prior to the development first being brought into use a verification report must be submitted to the Local planning authority demonstrating that the works have been carried out. The report shall provide verification that the remediation works have been carried out in accordance with the approved Method Statement. The development should not be brought into use until the verification report has been submitted to and approved in writing by the local planning authority.

e) In the event that it is proposed to import soil onto site in connection with the development the proposed soil shall be sampled at source such that a representative sample is obtained and analysed in a laboratory that is accredited under the MCERTS Chemical testing of Soil Scheme or another approved scheme the results of which shall be submitted to the local planning authority for consideration. Only the soil approved in writing by the local planning authority shall be used on site.

Reason
To ensure that risks from land contamination to the future users of the land and neighbouring land are minimised, together with those to controlled waters, property and ecological systems, and to ensure that the development can be carried out safely without unacceptable risks to workers, neighbours and other offsite receptors.

9. No development shall take place until the scheme of intrusive site investigations as detailed in the submitted Coal Mining Risk Assessment dated May 2015 has been undertaken. A report detailing the findings arising from the site investigation shall be submitted to the Local Planning Authority before any development commences. If any land instability issues are found resulting from, for example, past mining activity during the site investigation, a report specifying the measures to be taken to remediate the site to render it suitable for the development hereby permitted shall be submitted to and approved in writing by the Local Planning Authority. The site shall be remediated in accordance with the approved measures before development begins.

Reason
To ensure that risks from land instability and mining related hazards to the future users of the land and neighbouring land are minimised, and to ensure that the development can be carried out safely without unacceptable risks to workers, neighbours and the general public.

10. The hours of working on the construction of the development, and deliveries to/collection from the development site shall only take place between the hours of 7.30am and 6.00pm on Monday to Friday; 8.00am and 1.00pm on Saturday with no working taking place on Sundays, Bank and Public Holidays.

Reason
In the interests of residential amenity.

POSITIVE AND PROACTIVE STATEMENT

There were no problems for which the Local Planning Authority had to seek a solution in relation to this application.

NOTES TO APPLICANT

1. Pursuant to Sections 149 and 151 of the Highways Act 1980, steps shall be taken to ensure that mud or other extraneous material is not carried out of the site and deposited on the public highway. Should such deposits occur, it is the applicant's responsibility to ensure that all reasonable steps (e.g. street sweeping) are taken to maintain the roads in the vicinity of the site to a satisfactory level of cleanliness.

2. Due to the location of the site within a residential area the applicant should take all reasonably practicable steps to minimise noise and dust nuisance that may arise from activities on site. The applicant is also advised that material arising from the clearance of vegetation should not be burnt on site in order to prevent smoke nuisance being caused to neighbouring properties.

Officer: Steven Burgoyne
Signed:
Date: 14/08/2015

Checked By:
Date: 14/8/15

www.straightforwardco.co.uk

All titles, listed below, in the Straightforward Guides Series can be purchased online, using credit card or other forms of payment by going to www.straightfowardco.co.uk A discount of 25% per title is offered with online purchases.

Law

A Straightforward Guide to:

Guide to Your Human Rights and Civil Liberties

Consumer Rights

Guide to Business law

Guide to Public law

Mental Health and The Law

Bankruptcy Insolvency and the Law

Employment Law

Private Tenants Rights

Family law

Small Claims in the County Court

Contract law

Intellectual Property and the law

Divorce and the law

Leaseholders Rights

The Process of Conveyancing

Knowing Your Rights and Using the Courts

Producing your own Will

Housing Rights

Bailiffs and the law

Probate and The Law

Company law

What to Expect When You Go to Court

Give me Your Money-Guide to Effective Debt Collection

The Rights of Disabled Children

General titles

The Crime Writers casebook

Being a Detective

A Comprehensive Guide to Drink and Disorder

A Comprehensive Guide to Arrest and Detention

A Comprehensive Guide to Burglary and Robbery

Letting Property for Profit

Buying, Selling and Renting property

Buying a Home in England and France

Bookkeeping and Accounts for Small Business

Creative Writing

Freelance Writing

Writing your own Life Story

Writing Performance Poetry

Writing Romantic Fiction

Kate Walkers Guide to Writing Romance

Speech Writing

Teaching Your Child to Read and write

Creating a Successful Commercial Website

The Straightforward Business Plan

The Straightforward C.V.

Successful Public Speaking

Handling Bereavement

Individual and Personal Finance

**

The Completed House

End of a long but rewarding journey!